# GAMES FOR MATH

*Also by Peggy Kaye*

GAMES FOR READING

# GAMES FOR MATH

PLAYFUL WAYS TO HELP
YOUR CHILD LEARN MATH
From Kindergarten to Third Grade

WRITTEN BY *Peggy Kaye*

WITH ILLUSTRATIONS
BY THE AUTHOR

PANTHEON BOOKS
NEW YORK

All rights reserved under International and Pan-American Copyright Conventions. Published in the United States by Pantheon Books, a division of Random House, Inc., New York, and simultaneously in Canada by Random House of Canada Limited, Toronto.

Library of Congress Cataloging-in-Publication Data
Kaye, Peggy, 1948–
    Games for math.
    1. Games in mathematics education.    2. Mathematics
—Study and teaching (Elementary)    I. Title.
QA20.G35K39   1987          510′.7          87–45221
ISBN 0–394–54281–9
ISBN 0–394–75510–3 (pbk.)

Book design by Naomi Osnos.

Manufactured in the United States of America

B98765

# CONTENTS

Contents

vii

# ACKNOWLEDGMENTS

My best math teachers have been children. From James, age seven, I learned that children will love math—if they understand what they're doing. From Josh, age eight, I learned that when children make a spectacular mistake, there is usually a perfectly logical explanation. Nuj, who was seven and spoke Thai better than English, taught me that numbers know no language. Nina, age eight, taught me that when children understand the big ideas, the little ideas will take care of themselves. In teaching, there's no substitute for experience. I hope some of mine comes through to you in the pages that follow.

I am indebted to adults, also. To the teachers of New York City who helped me so much—Barbara Ridge and Gillian Ednie of the Walden School; Bernadette O'Brien, Paula Colon, Renee Lustig, Carlo Mitton, Sara Gay Klebanoff, and Donna Lynn Werner of P.S. 9; and Pedro Cordero, Shirley Maynard, and Gene Crichlow of St. Matthew's and St. Timothy's Daycare Center, all of whom allowed me into their classrooms to conduct experiments and make observations—many thanks. To Jerome Kaplan, whose expertise and patience were invaluable; to my excellent editor Sara Bershtel and her colleagues David Sternbach, Edward Cohen, David Frederickson, and Jennifer Dossin; to Marilyn Burns, Herbert Kohl, Allan Shedlin, Jr., Christine Stansell, and Sean Wilentz for the advice they gave; to Elizabeth Kaye, chef, and especially to Paul Berman, who magically fixed up sentences, paragraphs, in fact the whole book—thanks!

To my grandfather Joseph A. Daroff
1899–1987

# INTRODUCTION

## Play Games, by All Means

Every child is different when it comes to math. One child loves big numbers. Another child loves addition. A third sinks into despair at the sight of mathematical symbols—until suddenly, at age eight, he develops an unexpected math talent. In one way, though, all children are the same: they like math games. Put the right game in front of a child, explain the rules, and that child will eagerly play, happy and alert.

I've been teaching math to children privately and in classrooms for many years, and I've found that games are, from a teacher's point of view, wonderfully useful. Games put children in exactly the right frame of mind for learning difficult things. Children relax when they play—and they concentrate. They don't mind repeating certain facts or procedures over and over, if repetition is part of the game. Children throw themselves into playing games the way they never throw themselves into filling out workbook pages. And games can, if you select the right ones, help children learn almost everything they need to master in elementary math.

Games have another use, too. They offer a way for parents to get involved in their children's education. Common sense ought to tell us that parent involvement is a good idea; and if common sense doesn't tell us, there is objective evidence that does. One research study shows that when parents spend time every day helping their first-graders in math and reading, the children make significant academic gains. Another study shows that girls who excel in math tend to have parents—especially fathers—who encourage them in that direction. But how should parents get involved? What should they do? My answer is: play games. With the right games, parents don't have to know much about math. They don't have to worry about pushing or pressuring their children, and they needn't fear getting in the way of the school curriculum. All that parents have to do is propose a game to their child and start to play, in a spirit not really different from how they already play with their children.

There are fifty-six games in *Games for Math.* Several of them are well known to teachers. The rest I've made up over the years. In every case I've played these games with children from kindergarten to third grade, and have seen how effective they are. The games solidify the achievements of children who are already good at math, and they shore up children who need shoring up. They teach or reinforce many of the skills that a formal curriculum teaches, plus one skill that formal teaching sometimes leaves out —the skill of having fun with math, of thinking hard and enjoying it. If you play these games and your child learns only that hard mental effort can be fun, you will have taught something invaluable.

## Learning Math

Children have to learn an immense amount between kindergarten and third grade. They must learn to count, add, subtract, multiply, and divide. They must also learn about sizes and shapes, how to reason logically, how to work in systematic ways when solving problems, how to observe and analyze mathematical patterns, how to appreciate the beautiful designs of our number system. What a task!

To show how they accomplish all this, I will describe Julie, a perfectly ordinary child. Julie entered kindergarten already knowing something about numbers. She could recite from one to fourteen and was happy to demonstrate her ability anytime, anywhere.

She was eager to count higher, and by the time kindergarten was over, she could rattle off the numbers from one to one hundred. Julie's mastery, though, was not what adults may suppose. Often she made mistakes. She tried to count eight pennies, for instance, and came up with a total of ten. This was because she touched a penny for every number she recited, except that one time she forgot to touch a penny, and at the end she touched a penny that had already been counted, which left her with ten for an answer. Her mistakes were typical of young children. Counting more than five or six objects is harder than you think. And like most children, Julie was not really upset at her error. That's because young children see numbers in a different light than adults do.

One day I placed two rows of five pennies each in front of Julie:

I told her that one row of pennies belonged to her and the other row belonged to me. Then I asked, "Who has more pennies—you or me? Or do we both have the same amount?"

Julie answered, "We have the same."

I stretched out her pennies and left mine in place:

Again I asked, "Who has more—you or me? Or do we both have the same?"

Julie answered, "Now I have more."

I asked her to count her pennies. There were five. I asked her to count my pennies. There were five. Nevertheless she still believed she had more. Then I added a penny to my row:

I asked Julie about the two collections.

"Now they're the same," she said.

I asked her to count the pennies. This time she counted six for me and five for her. I said, "So, who has more—you or me? Or do we both have the same?"

"We have the same."

Very puzzling. It seems impossible that an intelligent child could count the pennies correctly and not understand that a change in count means a change in quantity. What do numbers mean if they don't tell us how many? To Julie, though, quantity meant something different than it does to you and me. Quantity to Julie was determined by looking at the pennies, not by counting. If one row looks bigger, it must contain more pennies. If the two rows look similar, they are the same—no matter what is shown by counting.

When adults discover this kind of thinking, they often react by telling children "the truth." But the real truth is this: adults can't say or do anything to change how children think about numbers. There was no way I could cajole or push Julie into adult-type thinking any more than Julie could push me into understanding quantum physics by tomorrow morning.

Julie had similarly odd ideas about shapes and sizes—that is, geometry and measurement. On entering kindergarten, she could identify and name a circle and a square. In the course of kindergarten, she learned to identify and name triangles and rectangles.

But when it came to measuring things, she was just as un-adult as when she tried to count objects. I placed two strips of paper in front of her like this:

She agreed that both strips were the same length. Then I moved one strip a little to the right, like this:

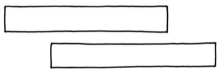

She thought the top strip was longer than the bottom. When I returned the strips to their original positions, Julie thought the lengths were again identical. That's how young children are. In their view, length can be transformed simply by moving strips from place to place.

Psychologists have a name for Julie's stage of mathematical thinking—pre-operational. At some point, usually between the ages of five and seven, children naturally and effortlessly enter the next stage, which is called the stage of concrete operations. Children in the stage of concrete operations know that even if you move the pennies, the number of pennies doesn't change, and even if you move the paper strips, the lengths of the strips remain the same. This knowledge is called the ability to conserve, meaning that quantity and size are conserved no matter how you move the pennies and strips. Children who conserve are ready to tackle all sorts of new subjects.

Julie developed the ability to conserve sometime in first grade; and with her expanded awareness, she was ready to take on the math curriculum of the first three grades. She learned to count high numbers, to count backwards, and to count by twos, threes, fives, and tens. She began her formal studies of addition and subtraction. By the end of third grade she had memorized about a hundred addition and subtraction problems.

She took up multiplication and division, at first using objects instead of symbols: dividing fifteen cookies among five friends, for instance. She accumulated a knowledge of geometry and measurement, fiddling with squares and triangles to learn about their special properties and getting to know exotic shapes like octagons, pentagons, and trapezoids. She studied centimeters and inches, liters, and quarts.

The hardest topic she studied was the peculiarities of our number system. She learned about the number ten and its role in the number system. She learned that the symbol 36 stands for thirty-six individual numbers—but also for three groups of ten and six ones. That was difficult to understand, but it was crucial. Learning to add and subtract big numbers depends on that understanding. All of higher mathematics depends on it. Of course, she wasn't ready yet for higher math. By the time she had learned everything I've mentioned, she was just in third grade.

## How to Use This Book

*Games for Math* has three parts. Parents of kindergarten children should pick from games in Part One, which are geared to a young child's brand of logic. For first-graders, you can pick any of the games in Part One and most of the games in Part Two. With second-graders you can play any game in the book, except for a few noted in Part Two. Third-graders can play any game in the book without exception. Each game is, in any case, labeled with suggested grade levels, so you needn't worry about remembering which part is for which grade. Of course, you shouldn't regard grade levels as a rigid rule. You have to observe how your particular child fares with a given game. Sometimes you'll notice that a game seems too easy. No problem. Playing a certain number of easy games is a good idea. Games that are too easy give children a chance to review. Children can make big educational gains by going over work they already know. Obviously, though, if you play too many easy games your child will lose interest. Pick a harder game, then, and see how the child reacts. If your child makes a lot of errors, you'll know the game is too hard, which is no good. Children learn more math and enjoy math more if they play games that are a little too easy rather than a little too hard. So if you end up with a game that's too hard, gently switch to something easier.

You don't have to play the games in any particular order. With grade levels in mind, you can skim through the book and start anywhere you want. You might play multiplication games before adding games, or geometry games before counting games—it makes no difference. One point does make a difference, however. You must pick games that seem like fun for you and your child. Refrain from games that don't entertain. I've tried to include activities for all tastes—board games, card games, dice games, story games, athletic games, guessing games, puzzles, even collages and cooking instructions—so it should be easy to find something that appeals.

The games are suited for every time schedule. There are talking games that take no preparation and can be played in a minute or two while you set the table or ride in a car. Other games take as long as fifteen minutes to get ready and half an hour to play. The game preparations are do-it-yourself, but the doing isn't hard. The equipment you'll need for most of these games is paper, pencil, index cards, paper clips, pennies, playing cards, dice, and an occasional bean or noodle.

*Games for Math* has some limitations. The book doesn't pretend to offer a complete math curriculum. Important topics like fractions aren't broached. Numerical estimation and certain other topics are addressed only in passing. Which is to say, the book is designed to help you back up a school math curriculum, but it can't replace that curriculum. Nor can the book tell you how to do your own private tutoring of a child who is floundering (though if your child does receive private instruction, at school or at home, you might want to discuss math games with the tutor and pick activities to reinforce what the tutor is doing).

Should you play games with your child every day or once a month? Should you play during school vacations and not during the school year? Should you play a favorite game over and over or try new games each time you play? Should you read all the games in the book or skim around until one of the games catches your fancy? Should you avoid a game that didn't pan out or try it again in a week, a month, or a year? The answer to each of these questions is: whatever seems agreeable to you and your child is what you should do.

The best advice on how you should play these games is—relax. Relaxing might mean changing the rules at the spur of the moment to better suit your child's personality. It might mean forgetting about a game halfway through because it stopped being fun. Relaxing means not worrying if your child seems to have trouble playing a particular game. Remember how it was when your child learned to walk? You weren't concerned about stumbles along the way. You weren't concerned because you knew that soon enough the child would not only walk, but run. You didn't pressure your child to take one more step, and you didn't worry if the baby next door walked a bit sooner than your baby. You patiently waited until your child was ready. Just so with math. Give the child a chance. Don't worry about the stumbles. Your child will learn to add and subtract and do all the other procedures in good time. Relax—all will be well.

You should remember that your child, too, has worries, which may be greater even than your own parental anxieties. First off, there's the issue of competition. Children find competitive games exciting, but also anxiety-provoking. Children want to win and are rarely good sports when they lose. Losing is painful. For some children, especially younger children, it is too painful, making competitive games a miserable experience for all involved. Does this mean you should avoid all competitive games or rig games to guarantee your child victory? In some few cases, yes. Generally, though, if you're sensitive to your child's tolerance for losing and stay within his frustration limits, things will go nicely. Occasionally you might rig a game in your child's favor, but not all the time.

Even during noncompetitive games—where winning isn't the goal—your child may get upset when he makes mistakes. That's because a child wants to perform perfectly in front of the most important people in his life, his mother and father. If your child finds mistakes hard to take, a little humor can be reassuring. Try saying, "It's OK to make mistakes—even I, your wonderful, terrific parent, sometimes make huge, jumbo, incredible boo-boos." Or say, "Thank goodness you made that mistake! If you never made any mistakes, I would never win a game. That wouldn't be fair." Or, "Finally, you made a mistake. Unless you make a few mistakes this game gets really boring."

Try picking out a game. Soon enough you'll know how your child feels about it. If you've given the game a fair chance, and the child frowns, put the game away. A game that's a drag, a bore, and a dud isn't worth wasting time that you and your child share. Somewhere on another page is a game that will cause a smile. That's the game for you. Does it call for rolling dice or picking cards or jumping around the room? Fine, do it. You and your child will enjoy each other; and your own pleasure will be that much greater, knowing that your child is slowly mastering mathematics.

# PART ONE

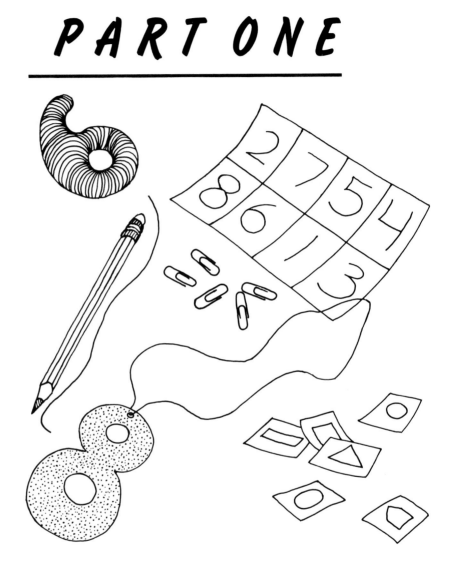

# chapter 1.

# Counting Counts

Spend a few hours in any kindergarten and you'll hear a lot of counting.

"One, two, three, four, five pretzels for snack."

"My doll's dress has three buttons."

"More crayons! One, two, three, four more crayons!"

"Look how tall my building is. I used ten blocks."

Sometimes the counting proceeds with no purpose at all, like a chant or a song. Children simply enjoy it. The endless repeating of numbers endlessly appeals to them.

It should appeal to you, too. Counting is extremely difficult. Children need to work at it for years. I asked a kindergartener named Sara to count a pile of pencils on a classroom table. She concentrated. She touched the pencils with her fingers. She whispered numbers, and then she said, "Fifteen." I was impressed, for in coming up with the right answer, Sara showed she had mastered three important concepts.

First, she had learned how to say the numbers one through fifteen. She knew to say them in a specific order, she didn't skip numbers, and she didn't add any extra numbers.

Second, Sara knew to assign one number, and only one number, to each and every pencil. Children don't learn this easily. The temptation to assign two numbers to a single pencil is very great. The temptation to disregard certain pencils and not count them is equally great. Even second- and third-graders, who know better, sometimes succumb to these temptations. In assigning one number, and only one, to each pencil, Sara had done something quite distinguished.

Third, she knew that the last number she counted was special. She understood that the number fifteen had a dual role. Fifteen was the name she assigned to the last pencil she counted. It was also the name for the entire collection of pencils. Children need no end of experience with numbers and counting before they can understand that fifteen can be the name for an individual pencil and, simultaneously, the name for a whole group.

The mistakes that children make show how difficult counting is. Many teachers tell stories like this. A three-year-old named David asked for some cookies. His mother helped David count out three from the cookie jar. After David ate one, his mother asked how many cookies were left. David said three. His mother was staggered: "How can you have three left when you just ate one?"

David responded, "I know I ate one, but I didn't eat two or three yet."

A sensible answer. In David's eyes, once *number three* became a cookie's name, *number three* it would always remain. He didn't understand that by changing the number of cookies on the plate, you change the name assigned to each individual cookie. That idea struck him as one more arbitrary and illogical rule devised by adults to befuddle children.

This chapter offers all sorts of games to give children practice with counting. There's a game with magazine pictures that helps children say numbers in the proper sequence. There's a picture collage that helps children discover the importance of using one and only one number for every object that is counted. There's a penny-tossing game ideal for active youngsters, a counting game you can play during bus rides, and a game that gets children touching toes, jumping, and hopping. There's also a game that helps children learn to write the numbers zero through ten. Browse and choose one. Try it out with your child. If the game is a hit with both of you, you're set. If it flops, try another.

# HOP, SKIP, TOUCH YOUR NOSE

**Y**ou can play HOP, SKIP, TOUCH YOUR NOSE anytime you have five minutes to spare—for instance, tonight after dinner. When dessert is done, turn to your child and say, "Before clearing the dishes tonight, sweetheart, will you please hop exactly six times?"

And so the action starts. Your child hops and counts, one number for each leap. You count along, to keep the numbers in line. After hop number six, you say, "Let's switch. This time you give the orders and I'll hop, skip, or touch my nose, depending on what you tell me to do. You have to tell me how often to do it. Go ahead, give me counting orders." Then come the orders, the action, and the counting out loud.

One or two rounds is enough. Then it's back to clearing the dishes.

How does this simple game help your child? Most kindergarteners can say the words *one, two, three, four* . . . But is that counting, or is it chanting, the way you say the alphabet? Chanting shows that a child knows the right sounds in the right order. But counting is about quantity. How many apples? How many pennies? Or in this case, how many hops, skips, or nose touchings. HOP, SKIP, TOUCH YOUR NOSE puts numbers in touch with actions. Playing this simple game gives counting a reality that children can feel from the tips of their toes to the tips of their noses.

**I**t was the summer before first grade and Dori was having trouble learning to count. She counted pretty well from one to ten, but the teen numbers were shaky and at twenty her counting ground to a halt. She needed help.

PENNY TOSS to the rescue. I took out a piece of paper, a pen, two pennies, and a box of paperclips (a box of toothpicks will do just as well). On the paper I drew a playing board divided into eight sections. I told Dori to choose eight numbers between one and ten, and I wrote one of these numbers in each section of the playing board.

**GRADES**
kindergarten and one

**MATERIALS**
paper
pen
paperclips or toothpicks
two pennies

We sat on the floor facing each other, with the playing board between us, right-side-up for Dori to read the numbers. I explained the rules: "We're going to take turns throwing pennies at the game board. You can start. You throw your penny at the board and try to land on one of the numbers. If your penny goes off the board or lands on a line, you throw again. When your penny lands on a number, you take that many clips from the paperclip box. Then I go. I throw my penny until I land on a number and can collect my paperclips. You throw a second time and collect more clips. I throw a second time and get more clips, too. Two throws for each player is all. Then we count our clips out loud to see who has the most. The one with the most is the winner."

Dori threw her penny. It rolled off to the side and she had to throw again. The penny landed on six. She picked out six paperclips and placed them in front of her in a neat pile. Next, I threw. I missed the board three times before my penny landed on the number four. I took four paperclips from the box, and Dori began to giggle.

"What's so funny?" I asked.

"I'm winning!" she said.

"Well, I still have a chance. Let's see what happens on the next throw."

Dori threw her penny and landed on eight. I threw mine and landed on six. We had to count to see who had the most paperclips. Dori was feeling so confident, she wanted to count her clips first. She counted to ten all right. After ten I recited the numbers along with her. That kept her counting on track. The game was riding on the count, so Dori was all attention. She seemed satisfied with her total of fourteen, but she wanted to know how many clips I had before she started celebrating. She listened very carefully as I counted, mouthing along when I got to eight, nine, ten. When I said ten, and Dori saw I'd come to the end of my clips, she clapped her hands in victory.

I suggested a rematch. In fact, we played two more rounds that day before going on to other activities. The next time Dori and I worked together, she asked if we could play PENNY TOSS. That was fine with me, but I had a suggestion to make. I proposed that this time, instead of throwing pennies two times apiece, we each get three throws. That would bring the counting into the twenties. Dori liked this suggestion, so we took out the PENNY TOSS game board and got down to play.

Dori and I played PENNY TOSS a lot over the summer. We threw four pennies eventually to get the numbers even higher. By Labor Day and the start of school, Dori was counting like a first-grade pro—up to thirty or forty anyway, which was good enough.

# NUMBER COLLAGE

**J**ill could recite the numbers one to ten—but she couldn't exactly count. I asked her to count a pile of ten beans and it went like this. She said "one" and moved a bean away. She said "two" and moved the second bean. At "three" she moved a third bean. But when she said "four," no bean moved. Five went well, but when she reached six, two beans slid across the table. Then she moved a bean but failed to say a number. Oh dear, that wasn't counting, that was confusion.

Carelessness was not the problem. Jill simply didn't understand the one-to-one relation between beans and numbers. She didn't understand that saying a number meant she had to move a bean, and moving a bean meant she had to say a number.

What to do for Jill? I explained NUMBER COLLAGE to her. She already knew about regular collages, having made quite a few in her day. A NUMBER COLLAGE is merely a collage with a set number of pictures. A four collage has four pictures. A seven collage has seven pictures. A fourteen collage has exactly fourteen.

I told Jill to pick a number. She picked twelve. Fine—we were making a twelve collage. I set out a magazine full of pictures, a scissors, a bottle of glue, a 14″ × 17″ sheet of paper and a large-sized colored marker. Jill's first job was to write a big number twelve in the middle of her paper. Then she searched the magazine for twelve pictures. Whenever she saw a picture she liked, she cut it out.

**MATERIALS**

magazines
scissors
glue
sheet of paper 14″ × 17″
or larger
pencil
colored marker

The pictures didn't need to have anything in common. Any old twelve would do. But size was a consideration: larger illustrations had to be trimmed to fit the paper. Some pictures were hopelessly big and had to be abandoned in favor of something more collage-size.

The main thing was, in order to keep track of how many pictures she'd snipped, Jill had to count. "Picture number one." "Picture number two." "Picture number three."

After the twelve were picked, clipped, and counted, we began to paste them on the paper. Pasting offered more opportunities for counting. I would ask: "How many pictures have you pasted?" "How many pictures left to go?"

When the pasting was done, we numbered each collage cutout. Alongside each picture, I wrote a number in light pencil marks. Then Jill traced over the number with a colored marker, which not only gave her more practice with the concept of one number per picture, but also provided a little practice in writing numbers.

NUMBER COLLAGE was perfect for Jill. The uniqueness of each picture helped her understand that with every picture goes a number all its own. The physical activity of cutting out and pasting made counting a concrete task—for children, the more concrete the activity the better. It takes time to make a NUMBER COLLAGE, and the slowness, too, was helpful. She couldn't understand numbers that whizzed past. The slow process of cutting and pasting was easier to follow. What's more, in our NUMBER COLLAGE Jill could actually see that the number twelve had two jobs. Twelve was the number Jill wrote next to the last picture she pasted on the page. It was also the big number she wrote to name the complete collage collection.

Jill worked hard on her collage, and I wanted to treat it with respect. I hung it in my workroom. Every time Jill came to see me she saw the collage displayed prominently on the wall. You'll want to treat your child's collages with respect, too. You can hang them in your house for all to admire, or you can give them as gifts to cooperative grandparents and other people who will be able to look at a bunch of magazine pictures glued to a piece of paper—and see a child's mind growing.

# THE HOW-MANY GAME

Here's a game you can play while riding on the bus, or waiting for food in a restaurant. Begin by asking your child to count all the people standing up in the bus. To answer, your child must scan the bus, assign a number to each standing person, and ignore everyone sitting down. If your child has trouble counting, help him out. After he discovers the number of standing people, ask him to count people with shopping bags, people with blue coats, people driving the bus, or people carrying umbrellas.

Once your child understands the game, turn the tables and let him ask counting questions for you to answer. To check your answers, your child will end up counting, too. Since every bus is full of things to count, you can stay busy right to your stop.

After you've been counting for a while, you can add one more type of question. In this bus, how many singing penguins do you see? How many flying blue rabbits? How many talking pocketbooks? How many people sitting next to rhinoceroses? The answer to all such foolishness is zero.

True, these questions are funny, but are they educational? Yes, indeed. Zero is a hard number to understand. Children have tangible experiences with all other numbers. You can pet one cat, hold two eggs, and collect three pebbles. But you can't pet, hold, or collect zero cats, zero eggs, or zero pebbles. Zero is usually left out of counting. You don't count carrots by saying, "Zero, one, two, three, four." Since children's experience with zero is limited, they don't understand this number very well. Ask your child to count all the polka-dotted elephants in a bus, however, and he may laugh at your silliness, but he must also come up with a number. If the number isn't zero, either your child needs more practice, or the people in your bus have a problem.

# THREE KINDS OF FANCY NUMBERS

**GRADES**

**kindergarten and one**

**MATERIALS**

**a large piece of drawing paper (or oaktag or poster board)**
**colored markers**
**pencil**
**play dough**
**colored index cards**
**white glue**
**glitter**
**scissors**
**newspaper**
**yarn**

Sometimes children can say the numbers out loud, but can't write them down correctly. They do it backwards so the numbers come out Ɛ or Ƨ or ᔭ .

Parents worry when they see this. But the problem can usually be solved with practice. Here are three ways:

## Poster Numbers

Get a piece of paper, oaktag, or poster board 11″ × 14″ or larger and write a huge number five that fills up the page. Now give this paper to your child so he can decorate the number with colors, stars, and stripes.

When this giant number is finished, what can you do with it? Put it on the floor and ask your child to tip-toe over it, letting his feet imitate the movement his hand uses to draw the number. Call the number a tightrope wire and dare your child to stay on line from beginning to end. Can he do it? Shout hooray, bravo, encore.

Then make a new poster with a different number. Four is the only number drawn with two separate strokes. Your child can walk this number correctly if he jumps from one stroke to the next.

When you're through drawing, decorating, and walking all over these posters for the day hang them up on the wall. Soon you'll have a display of 0, 1, 2, 3, 4, 5, 6, 7, 8, and 9, all the symbols needed to make every number in this world.

## Play Dough Numbers

Make numbers with play dough? Sure, it's easy to do. Just help your child roll play dough coils with his favorite colors and assist him in twisting the coils into number shapes.

When you're finished play-doughing for the day, you don't have to mush up the numbers. You can save them. And when they're dry, brush watered-down white glue over them. The glue patches up broken spots and helps prevent the numbers from crumbling.

When the glue dries, try playing the following game. Blindfold your child. Then place a number in front of him. You can put the number right side up: or reverse things by flipping the number over:

Then place your child's hand on top of the number. He has to tell, just by touching, whether the number is facing the right or wrong direction. After he makes a determination, slip off the blindfold so he can check the accuracy of his touch.

When you're done with this game, you might use the numbers as paper weights.

### Glitter Numbers

Take an index card and lightly outline (in bubble drawing) one of the numbers zero through nine.

Give the card to your child and let him cut out the number. If your child finds the cutting too difficult, do it for him. After the number is cut out, protect your table with newspaper and bring out the glue. Have your child smear glue over the top side of the number. Usually this causes the index card paper to curl up. If you gently press the number down, it will lie flat again. Quickly now, while the glue is wet, sprinkle the number with glitter.

What can you do with this charmingly glittery number? You can tack it on a bulletin board. Or mount it on a different-colored index card making a lovely number picture. Better yet, you can punch a hole in the number, thread it with yarn, and wear it as a glitter-number necklace.

Does FANCY NUMBERS really help children? Usually when youngsters write numbers, they use a pencil and draw thin little lines. Those lines aren't particularly memorable to beginners. But with FANCY NUMBERS, the giant size, the 3-D form, and the glittery cutout bring numbers to life. Once a child walks on a 5, holds it in his hand, or wears it as a necklace, he's less likely to draw it backwards with a pencil.

Of course, you have to be patient. Anyone can make an occasional mistake.

# chapter 2.

# Thoughts to Grow On

The Swiss psychologist Jean Piaget spent his life studying how children learn to think. He discovered, through an ingenious series of experiments, the exact ways in which children's ideas differ from adults—about numbers, for instance. It's easy to replicate some of Piaget's work. I did it with Sam, a boy just starting first grade.

I put five paper cups in a row in front of Sam. I showed him a collection of daisies. Then I asked him to place one daisy in each cup. Sam did this with ease.

I asked him if there were the same amount of cups as daisies.

"They're the same," he replied.

I took the daisies out of the cups and spread them out on the table. "Are there the same amount of cups and flowers now?" I asked.

"No," answered Sam. "There are more flowers."

"Why do you think so?" I asked.

"Because the flowers go all the way to here," Sam answered as he pointed to the last daisy in the row. "But the cups don't go as far."

"Did I add any flowers or take any away?" I asked.

"No," he answered.

"Did I add any cups or take cups away?"

"No," said Sam.

"But there are more flowers now than cups?"

"Sure," he said.

I asked Sam to count the flowers and to count the cups. There were five of each. He was amazed.

"Well, then," he said with some discomfort, "they must be the same." This showed that Sam was influenced by counting (not all children are). But the influence was temporary. I took the cups and spread them out, while I pushed the flowers closer together.

Sam instantly reported that now there were more cups. When I added an extra flower to make the row of cups and the row of flowers the same length, Sam was sure that six flowers were equal to five cups.

All children, without exception, spend many years sharing Sam's viewpoint. You had these misconceptions when you were young. So did Einstein.

Generally between the ages of five and seven, children's ideas about numbers undergo a transformation. Youngsters realize you can change floral arrangements without affecting the number relation between flowers and cups. That was the case with Sam. Six months after our first experiment, we had a replay. This time Sam answered all the questions about cups and flowers correctly.

What happened in just six months to change Sam's analysis of cups and flowers? It wasn't that Sam could now count more accurately. Counting was never his problem. What changed was Sam's mathematical thinking. At the beginning Sam thought flowers and cups were different in number when they looked different. Now he understood that if you don't add or remove any cups or flowers, you can't change the number of cups or flowers on the table—no matter how things look. He understood that logic counts and looks do not. His reasoning had developed. He had left one stage of mental development behind, and advanced to the next.

Educators sometimes like to think that children can be hurried along from one mental stage to the next. But it can't be done. Trying to hurry this development is like trying to make children grow quicker. It won't happen. Piaget discovered this in his experiments, and educational studies have consistently confirmed it. That doesn't mean that teachers and parents have no role to play. If you want your child to grow, you give him good food—and wait patiently. If you want your child's reasoning to grow, you should give him good intellectual food—mixed with patience.

In this chapter you'll find three different types of brain food to whet your child's intellectual appetite. There's an activity that helps children think logically while sorting groceries. There's a game involving hops, skips, and jumps organized by size. There are activities to help children observe patterns.

None of these activities has to do with numbers, directly. None of the activities will *teach* your child that five flowers and five cups are equal in number no matter how they are arranged. That knowledge will come in its own good time. But the activities encourage logical thought. They give children the habit of thinking hard.

# THE ER GAME

The ER GAME is a jumping, hopping, clapping game. Don't worry, it's only the child who jumps, hops, and claps. Your job is to bark out orders.

"Today we are playing the ER GAME. To play you must follow my exact instructions. First, I want you to jump. Go ahead, jump."

"Oh, that was super. Now I want you to jump again—only higher."

"Good for you, that was really up there. This time, I want you to jump even higher. Do you think you can do that? Do you think you can jump even higher? Go on, do it."

Once your child has jumped high, higher, and even higher, switch commands. Tell him to step, step longer, step longer yet. Hop, hop smaller, hop smaller yet. Clap, clap louder, clap louder yet. Shout, shout quieter, shout quieter yet.

The ER GAME is fun, but five minutes is the most you'll want to play at a stretch. Luckily, stray five minutes come often. You can hum and hum louder or clap and clap faster while making beds or chopping vegetables for dinner. You can skip and skip smaller or leap and leap bigger as you walk to the bus or the store.

The skipping and leaping helps your child learn to put things in order according to size. With each leap your child must make a decision. He must decide how big a bigger leap must be, compared to his last leap, which was bigger than his leap before. Numbers are another way to put things in size order. Five is bigger than four, four is bigger than three. Six is smaller than seven, seven is smaller than eight. There are no numbers in the ER GAME. Even so, getting children to think about sizes of jumps and hops, about loudness of claps and shouts, prepares them to understand the big, bigger, and bigger-yet relationship between five, six, and seven. It helps them understand the small, smaller, smaller-yet relationship between eight, seven and six.

# KITCHEN CALCULUS

**Y**ou've just returned from the supermarket. Unpacking your bags, you put a big pile of groceries on the kitchen table. Now you call your child into the kitchen.

"Today is your lucky day. You're going to help me put the groceries away. Here's how. I'm going to stand by the refrigerator while you hand me everything that gets stored in the cold. If you make a mistake and hand me something that belongs in the cabinet or in the closet, I'll put it back on the table. When you think you've found everything that belongs in the refrigerator, I'll look at the table and see if you missed anything.

"Then we can start putting away things that go in the closet, then things that go in the cabinet, then things that go on the kitchen counter. Soon the table will be empty, but the refrigerator, the cabinet, and the closet will be full."

Important thinking goes on whenever you play this game. As your child selects groceries, he makes lots of decisions. Does broccoli belong in the refrigerator? How about mayonnaise? What about tuna fish? By answering these questions, your child is forming a foods-that-belong-in-the-refrigerator group. Some of these foods are similar, like orange juice and apple juice. Some are very different, like ice cream and chicken. But ice cream and chicken have something in common when it comes to classifying groceries. KITCHEN CALCULUS helps your child understand abstract classifications like things-that-belong-in-the-refrigerator. Numbers are merely another system of abstract classification. The child who understands the difference between things-that-go-in-the-refrigerator and things-that-go-in-the-cabinet will have a leg up on understanding that three books, three sponges, and three pencils go in the number-is-three group, while four books, four sponges, and four pencils go in the number-is-four group.

Needless to say, the game might also get your child in the habit of putting away the groceries, which couldn't hurt.

**GRADES**
kindergarten and one

**MATERIALS**
a bag of groceries

# OODLES OF NOODLES

**GRADES**

kindergarten and one

**MATERIALS**

a box of ziti noodles
colored construction paper
glue
pencil or pen

How many different ways can you arrange six macaroni noodles on a page? I handed Benjamin a piece of colored construction paper, some glue, and a box of ziti noodles, and told him to get to work. The rules were: he could use six zitis in each design; each ziti had to touch at least one other ziti in the design; each design had to be unique.

Benjamin began arranging. When he had a design he liked, we glued it to the colored paper. I labelled the design with a six and drew a circle to keep these zitis from getting mixed up with any others.

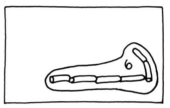

Then Benjamin counted out six more noodles and went back to work. He designed, we glued, I labelled and circled, and we were on to the third design, and then a fourth. There were oodles of noodles on that page. We counted. Benjamin had created six different six-ziti designs. He might have come up with more, but his paper was full.

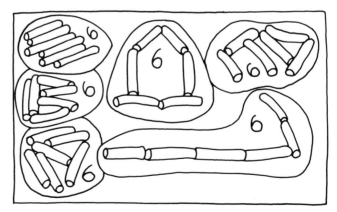

Benjamin was thrilled. He wanted to do it again. So I gave him a fresh sheet of paper and suggested that this time he make an eight-ziti collage.

What cunning logic lies behind OODLES OF NOODLES? It is this: Benjamin was like every child—he misunderstood numbers. In his heart of hearts, he believed that shape determines quantity. He thought that

contained a smaller amount than:

even though both contain, in fact, six zitis. Only time and experience would help Benjamin discover the truth about numbers—that arrangement doesn't affect quantity. OODLES OF NOODLES provided Benjamin with experience. By making six-ziti collages, he got to see six in this way, that way, and the other way. In his own unrushable time, and given many experiences like OODLES OF NOODLES, Benjamin's thinking about numbers would inevitably undergo a drastic transformation.

Piaget tells a story of a friend, a mathematician, who remembered the exact moment his thinking about numbers changed. He was five years old, playing with pebbles on the sidewalk. He spread the pebbles in a row and counted them. There were ten. He arranged the pebbles in a circle and counted them. There were ten. How surprising: the pebbles in a line looked different from pebbles in a circle. He arranged the pebbles in a triangle. To his delight and amazement, when he counted the pebbles, there were exactly ten. Aha, he thought, no matter how I arrange them, there are always ten pebbles. Ten pebbles this way, ten pebbles that way, ten pebbles the other way. The pleasure of this discovery stayed with the mathematician all his life.

# CLAP CLAP
# BEEP BEEP

*T*he number world is filled with patterns. Here's a sophisticated example that calls for multiplying 37 by multiples of 3:

$$37 \times 3 = 111$$
$$37 \times 6 = 222$$
$$37 \times 9 = 333$$
$$37 \times 12 = 444$$

Can you predict the answer to $37 \times 15$? Good mathematicians, whether they are adults or children, love discovering and fiddling around with patterns. They delight in the beautiful predictability of numbers.

You can help your child develop this skill by playing find-the-pattern games. These games needn't be number games. In fact, it's better if they are not. The absence of numbers leaves children free to concentrate on patterns.

CLAP CLAP BEEP BEEP is an excellent game for beginning pattern-makers. You start by clapping in a regular pattern: LOUD, soft, LOUD, soft, LOUD, soft. Give your child a chance to listen, and when he's got the beat, he can join in. Now, in unison, you and your child clap LOUD, soft, LOUD, soft.

Follow up this clapping pattern with a vocal arrangement. Tell your child to listen carefully while you begin a chorus of, "beep beep, honk honk, beep beep, honk honk." Your child can beep and honk along, as soon as he's figured out your song.

For your next pattern, slap your knees, hum a tune, slap your knees, hum a tune. These simple, easy-to-follow patterns are perfect for beginners. As your child develops a savvy for patterns, try more complicated combinations: sit, stand, jump, jump; sit, stand, jump, jump; sit, stand, jump, jump. Then go for some noise: bang the table, shout "beep beep," stamp your feet, whistle; bang the table, shout "beep beep," stamp your feet, whistle; bang the table, shout "beep beep," stamp your feet, whistle.

Until now, you and your child have performed duets. The time has come to launch your child on a solo act. First, you establish a pattern. Second, let your child join in. Third, you drop out, leaving your child to repeat the pattern on his own.

When your child is accustomed to keeping up a pattern on his own, you're ready for a grand finale. Set your child up with a pattern. Beep, honk, beep, honk, beep, honk will do nicely. When he's going great guns on his own, you enter in with a new and different combination of sounds: squish, squash, squish, squash, squish, squash. You keep squishing and squashing while your child beeps and honks. A great moment in family history!

Your child will probably want to invent his own patterns rather than sticking to ones you make up. Go ahead and let him. But be prepared: children get so carried away with honks and jumps, they forget that pattens must repeat in a regular fashion. Eventually children get the idea—but it takes a while.

# WHAT'S NEXT?

**I** wiggled my fingers, stamped my feet, wiggled my fingers, stamped my feet, wiggled my fingers, stamped my feet, and then asked Dionne, "What comes next?"

At first my young playmate looked at me like I was crazy. But she proved she understood the game when she started waving her fingers in the air.

"This comes next!" she shouted, wiggling like mad.

"Excellent. Here's another." I stuck out my tongue, shook my head, stuck out my tongue, shook my head, stuck out my tongue. Dionne was still giggling when I stopped and asked, "What's next?"

Dionne shook her head.

"Ooh, you are good at this game," I said. "I'll have to make things harder if I want to trick you. See if you can figure out this puzzle." I slapped my knees, slapped my shoulders, shook my head, wiggled my fingers, slapped my knees, slapped my shoulders, shook my head, wiggled my fingers, slapped my knees— and then I stopped. "What's next?"

Dionne began to wiggle her fingers.

"Finally, I tricked you. I had to work hard to do that. But I'll give you a second chance. Here's the pattern again."

I repeated and repeated my actions until I saw a light of recognition in Dionne's eyes. Then I stopped and asked, "What's next?" This time she got it right.

That was enough for one day. A week later, Dionne and I played the game again. Because I was feeling low on energy, I decided to forget action patterns, and played a talking WHAT'S NEXT? instead.

I started with the feline pattern, "tiger tiger, cat cat, tiger tiger, cat cat." When I stopped after "cat cat" Dionne instantly growled "tiger tiger."

My second pattern was a little harder: "truck, bus bus, bike, truck, bus bus, bike," but Dionne sped through it.

I made the next pattern—a name pattern—really tough: "Peggy, Dionne, Peggy, Paul, Peggy, Dionne, Peggy, Paul." Unfortunately, this pattern was too hard for Dionne. I gave her a couple of chances to figure it out, but she was stumped.

Naturally, I made the next pattern slightly easier to predict: "spin spin, top, spin spin, top, spin spin, top."

Dionne and I played WHAT'S NEXT? often during the school year—always for about five minutes at a go. Some days I stuck to word patterns, other days I mixed words and actions. On rare days, I created jumping, skipping, hopping, and twisting patterns. On those days, a pattern workout for Dionne became an aerobic workout for me.

# chapter 3.

# Size and Shape

Here are two strips of paper the same length:

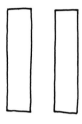

If I move one paper down a little bit, are the two strips still the same length?

Here's a straight strip of paper:

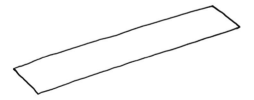

If I bend it into a zigzag shape, is the paper still the same length?

If you are a grown-up, you'll answer yes to both these questions. A child younger than seven, however, will probably answer no.

When young children look at the two nonaligned strips of paper, they only pay attention to one end at a time. They see that one paper is longer at the top, but they don't simultaneously consider that the other paper is longer at the bottom. Children don't reason that if the two strips were the same length a moment ago, and all you did was move one strip, the two strips must still be the same.

When young children watch you make a zigzag strip from a flat strip, they notice that the zigzag strip doesn't stretch as far as the flat strip did. They assume, therefore, that when the paper is zigzagged it's a different length than when it's flat. They believe, in short, that size changes with shape.

They absolutely believe this. You may badger them into saying the right answer, but their conception of size and shape won't, in fact, change. There are, however, aspects of size and shape that young children can understand—and that will prepare them for more formal measurement learning later on. That's what the activities in this chapter are about. In one activity you and your child will make ribbon measurements all over your house. In another, you and your child will measure your waists, fingers, toes, and noses. Children can also learn to measure capacity. The game HOW MUCH DOES IT HOLD? teaches this with a bit of splashing around. There are also games to help children learn about shapes like squares, triangles, and rectangles.

Chapter Six offers additional activities in measurement and geometry—activities appropriate for older children. When your child is ready to grapple with advanced aspects of size and shape, you can turn to page 89 and plunge in.

# STRING TIME

**S**TRING TIME is a measuring contest with sure-fire child appeal. When Josh, an energetic kindergartener, was feeling bored with other math activities, we livened things up by playing this game.

I introduced STRING TIME by asking Josh to look at a pencil lying on the table. I told him to study its length from point to eraser. I told him to study for as long as he wanted, and in any way he liked, but not to touch the pencil—not even once. When Josh announced he'd looked enough, I took a ball of string and a scissors from my drawer. I told him I'd been studying too. Now I would try and cut a piece of string the same length as the pencil. I knew I wouldn't get the exact length, but I wanted to come as close as possible. After explaining this to Josh, I pulled some string from the ball. Using my finger, I pointed to different places on the string, trying to get the right length. When I thought I had the proper distance staked out, I cut.

Next Josh got his turn to cut a string as close as possible to the pencil length. I warned him to work carefully, because when he was done we would compare our strings with the pencil. Whoever came closest would win the contest. But there was one catch. The string couldn't be longer than the pencil. Whoever cut the string longer than the pencil automatically lost.

With these instructions in mind, Josh fiddled with the string for a bit; then he cut. The moment of truth was upon us. We laid the two pieces of string alongside the pencil, using the pencil point as an end line. Then we judged.

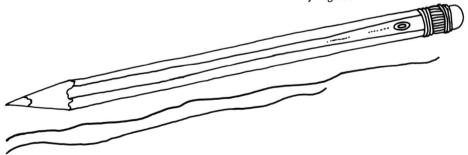

Neither of us had the perfect length. Both strings were too short. But Josh's string came closer. He won.

Josh was eager to measure something else. We set our sights on the front side of a table. We studied the length, we cut strings, and we compared. Josh's string was too long, while mine was too short. Josh knew the bitter truth—he automatically lost. Before going on to measure something else in the room, Josh asked what would happen if both strings were too long. I told him in that case, we'd both lose.

Josh and I never ran out of things to measure. We measured the height and width of a tissue box, the width of a dictionary, the length of a TV screen, the circumference of a flower pot.

The idea was gradually to prove to Josh he could measure anything in sight—big things and little things, the movable and the immovable. Because we measured heights, widths, depths, diagonals, and circumferences, Josh came to realize he could measure objects from many different angles and directions. That is a novel thought for young children, and it accounts for some of the fun in STRING TIME. Finally, the game got Josh in the habit of thinking hard about an abstract quality—size.

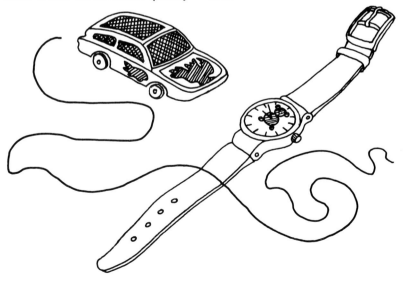

# RIBBON ME

**GRADES**

kindergarten and one

**MATERIALS**

a reel of ribbon

scissors

large piece of oaktag or
poster board

glue

pencil

**R**IBBON ME gives a young child the chance to measure the most interesting person in the world—himself. Many parents know the delight children take in charting their growth. That's why, in any number of homes, a special wall is covered with marks indicating children's height from month to month or year to year. But why limit the measure of a child to his height?

Your child might like to know the length of his arm, leg, or nose, the circumference of his head, waist, or ankle, or the span of his palm. Instead of using marks on a wall to record these measurements, try using lengths of ribbons and preserving them on a RIBBON ME chart. Begin the chart by measuring your child's foot. Take a reel of ribbon, stretch one end out so that it touches your child's toes, and have your child hold this end in place. Then pull the ribbon so that it runs along your child's foot to his heel. The child will need to use both hands—one to hold the ribbon at his toe and the other to hold the ribbon at his heel. While your child holds the ribbon in place, you cut.

Glue the foot ribbon to the poster board, which is your chart.

Next try measuring your child's waist. The child holds one end of the ribbon at his belly button, then pulls the ribbon around until it reaches the starting point. You make the snip. That's the waist ribbon, and up on the chart it goes.

It's fun to use these ribbons to compare body parts. You will discover that the length of your foot is identical to the distance between the inside bend of your elbow and your wrist, which is a little known fact of human anatomy. In order to make honest comparisons between body parts, you have to draw a base line on the chart. The bottom of your foot ribbon will do as the starting point.

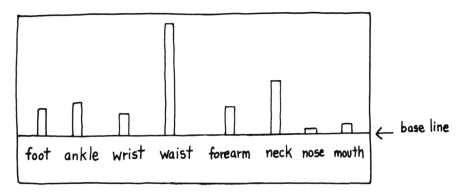

What happens if there is no base line? You might place a wrist ribbon alongside an ankle ribbon like this:

That's when many children reason that the wrist ribbon is longer. It does, after all, go higher up. Give the ribbons a common starting point, and children aren't so easily fooled.

# HOW MUCH DOES IT HOLD?

**GRADES**

kindergarten and one

**MATERIALS**

plastic or metal cups
jars
pans
bottles
all sizes and sorts of
unbreakable containers
newspaper
optional: a plastic or metal funnel

**P**our some orange juice into a tall thin glass and show it to a young child. Then have the child watch as you pour the same juice into a short wide glass.

Does one glass have more juice or do they both have the same? Until a child is between five and seven years old, he believes the two containers hold different quantities of juice. Most children think the taller glass holds more because the juice goes up so high. Some children think the short glass has more, because the juice stretches out. No wonder young children squabble over who got more juice for lunch!

You can't teach your child that the two glasses hold the same amount of juice. You can, however, introduce your child to the idea of measuring capacity. That way, you get him thinking about HOW MUCH DOES IT HOLD?

HOW MUCH DOES IT HOLD? is a water game, and you should only play on a day you're willing to mop the floor. When that day comes, collect a variety of unbreakable containers in different sizes—for instance, plastic cups, paper cups, an empty soda bottle, a plastic soup bowl, plastic refrigerator containers of different sizes. A plastic funnel is helpful, though not necessary. Cover the floor with newspapers or some such thing to soak up the overflows. Fill a sink or the bathtub with water. Then challenge your child to a HOW MUCH DOES IT HOLD? contest. Start by spotlighting two different containers, and make a guess.

"I think this soup bowl will hold five paper cups full of water. Do you think I'm right? Or am I wrong? How many cups do *you* think the bowl will hold?"

Give your child a chance to make an estimate, then start measuring. Fill a paper cup with water. Pour the water into the soup bowl. Fill the cup again and pour into the soup bowl. Keep going until the soup bowl is full. When I did this, it took four paper cups of water to fill one soup bowl.

How many soup bowls of water can your refrigerator ice-tray hold? How many plastic cupfuls of water do you think a bread pan can hold? How many paper cups of water will it take to fill a soda bottle? Go ahead, take your best guess.

Both you and your child should make estimates before you pour. But don't be surprised if your child is way off. Children are bound to be swayed by visual impressions you have long since learned to ignore. Eventually your child, too, will learn. Meanwhile you're helping him begin to think about two mathematical abstractions, volume and capacity.

When is the game over? When it's no longer fun, or when there's more water on the floor than in the sink.

# SHAPE LOTTO

**GRADES**

kindergarten and one

**MATERIALS**

oaktag or poster board
index cards
scissors
crayons or colored pencils

**S**usan couldn't remember the names of simple shapes like rectangles, triangles, and squares. The shapes seemed strange to her. She needed to get used to them, and to saying their names. No problem—we played SHAPE LOTTO.

First, though, I had to spend about fifteen minutes making game boards and cards. I started with two boards, one for each of us. Oaktag or poster board is best, though any kind of paper will do in a pinch.

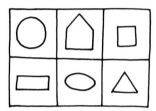
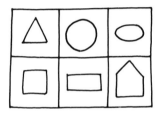

As you see, each board has a rectangle, a square, a triangle, a circle, an oval, and a pentagon, but in varying positions.

Next I took seven index cards and cut them in half. On two of these cards I wrote *sorry*. On the other twelve cards I drew circles, ovals, pentagons, triangles, rectangles, and squares, two cards for each.

We were ready to play. I shuffled the cards and spread them out on a table top, face down. Susan took one playing board and I took the other. I explained the rules:

"To win this game, you have to cover all the shapes on your

playing board before I cover all my shapes. Here's how this works. I pick a card from the table. If I'm lucky, I can use this card to cover one of my shapes. Look—I got a triangle. That means I can cover up the triangle on my board with the triangle game card. Now it's your turn to pick."

"I got a circle," Susan said, after picking a card.

"Good for you, you can use that card to cover the circle on your board. It's my turn. Oh no, I got *sorry*. I can't cover up anything. I'm going to put the *sorry* card back on the table and mix it up with the other cards. Your turn."

"I got a . . . what's this called?"

Well! That was the point of the game. I got to say, "It's a rectangle. You're really lucky because now you've covered up your rectangle and your circle."

Then it was my turn. I said, "I hope I get a rectangle like you did. Come on, rectangle. Come on, rectangle. Let me find you, you old hidden rectangle. Hey, I got it. We're tied."

"I hope I get a square. I hope I get a square. I hope I get a square," chanted Susan as she passed her hands over the cards.

The game was getting to her. She picked. "I got a circle. But my circle's already covered."

"That means you must put the circle back with the other cards, and it's my turn. Oh boy, I hope I get that circle this time. Come here, you little circle. Come here. Phooey—another *sorry* card. Ugh. I'll hide it back with the other cards. Your turn. What card do you want to get?"

"I want this one," said Susan, pointing to one of the shapes on her board.

"You want the pentagon," I added. "Pentagons are shapes with five sides. See, you can count the sides: one, two, three, four, five."

"I want a pentagon. I want a pentagon," said Susan as she picked a card. "I got it! I got the penta . . . penta . . ."

"Pentagon," I contributed. "Your pentagon, your circle, and your rectangle are covered up. All you have left is your triangle, your oval, and your square."

It took five more rounds before Susan filled up her board and won.

"Can we play this again some day?" Susan asked.

"Sure," I answered. "I'll put the game on a shelf and I promise we can play again next week. How's that sound?"

"Great," Susan said.

I thought it was great, too. There was no amount of straight-forward teaching that could have got Susan to focus so intently on shapes and their names.

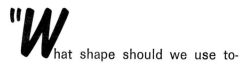

**"W**hat shape should we use to-day?" I asked.

"Triangles," answered Sophie.

"You got it," I said. "What colors would you like?"

"I want yellow and blue and red."

"I like those colors too," I said. I took a sheet of yellow construction paper and drew five triangles on it. Each triangle was a slightly different size and shape.

I handed this paper to Sophie and told her to cut out the triangles. While she was cutting, I drew triangles on the blue and red sheets of paper. Then I was ready to help Sophie cut. Ten minutes later we had a colorful collection of seventeen triangles.

We were ready to make triangle collages. I gave Sophie a large sheet of paper and a jar of glue. I took a piece of paper for myself. We started positioning triangles on our respective papers and gluing them in place. When we finished, we had two lovely triangle collages to hang on the wall.

**G R A D E S**

kindergarten and one

**M A T E R I A L S**

colored construction paper
pencil
two scissors
glue
a large sheet of drawing paper,
approximately 14" × 17"

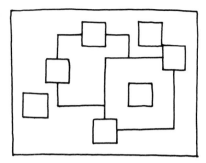

These were two of many shape collages Sophie and I made during the school year. Each collage was different, because we used different shapes each time. We made square collages, oval collages, rectangle collages, square-and-circle collages, square-rectangle-triangle-and-hexagon collages. Sometimes each collage cutout was unique. But we could also make a series of identical cutouts by tracing a single shape over and over on the construction paper.

What did Sophie learn while creating these shapely art works?

First, she learned the names of geometric shapes. By the end of the school year, Sophie knew the names for circles, triangles, rectangles, and squares. She knew, although sometimes she forgot, the names for ovals, octagons, hexagons, and pentagons. What's more, she knew that all four-sided shapes, including squares and rectangles, can be called quadrilaterals.

Second, Sophie developed multifaceted ideas about geometric forms. Before we made collages, Sophie recognized these shapes as triangles:

But she didn't know what to call these:

Third, Sophie noticed some interesting peculiarities about shapes. She noticed that two equal-sized squares glued together can make a rectangle:

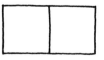

She also discovered that two equal-sized triangles glued together can make a rectangle:

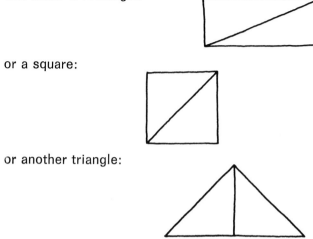

or a square:

or another triangle:

She found that when you overlap squares, you get triangles.

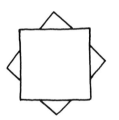

Sometimes she made these discoveries herself. Other times, I glued the shapes in special ways and showed Sophie what I'd done.

At home, it will take half an hour from start to cleanup to make a SHAPE COLLAGE. It's a perfect activity for rainy days. Two children can work on it together. You might consider SHAPE COLLAGES the next time your child has a friend over to play. If your child has trouble cutting out the shapes, you must take on this job for yourself. You can cut out the shapes Friday night while watching TV, and use them for a collage Saturday before lunch.

# PART TWO

# chapter 4.

# Fancy Counting

First-graders generally learn to count, read, and write from one to one hundred. Second-graders learn to count, write, and read from one up to nine hundred ninety-nine. Third-graders get up to a million. But in making this gradual progress, children also begin picking up fancier and fancier ways to count, which is important when the time comes for addition, subtraction, and multiplication.

Sometime in first or second grade, most children discover they can begin counting from numbers other than one. To count starting from five, say, or six—that's a big step in a child's development. Soon enough, children discover how useful this way of counting can be, especially when it's time for addition. Consider the problem, five plus three. Children initially solve this problem by counting from one to five. Then they count three more numbers to get the answer. Eventually, though, comes an intellectual leap. Children discover they can start counting at five, count three more numbers, and answer the problem. That's quicker and easier.

Children tackle a second counting technique when they learn to count backwards. Counting backwards is important for subtraction. Seven minus four—that's six, five, four, three. Most first-graders know the missile countdown: ten, nine, eight, seven, six, five, four, three, two, one, blast off. Yet many of these same children can't go backwards from seven or six. It's even more difficult for children to count backwards from numbers greater than ten. Come to think of it, I myself have to stare into space a few seconds before counting backwards from one hundred million.

Skip-counting—counting by twos, threes, fives, or sixes—is a third way to count. Skip-counting is a skill that develops slowly. When you've learned it, you've learned the multiplication table. Four, eight, twelve, sixteen, twenty—you're halfway through the fours table.

In this chapter there are bean games, tricky contests, and hopping games, all designed to help first- through third-graders learn to count large numbers, start counting from numbers other than one, count backwards, and skip count. There's an estimating game to help children develop good number sense while learning to count. There's a find-the-hidden-penny game to help children count ordinal numbers: first, second, third, fourth, fifth. Some evening while watching TV, or waiting for floor wax to dry, skim through *Fancy Counting* and find a game that suits your fancy. Then, when the moment seems right, play it with your child. The game may go badly. If so, stop playing and pick some other activity. When it comes to games, what doesn't amuse shouldn't be used.

# FILL THE SPOON

**GRADES**

one and two

**MATERIALS**

two teaspoons
two tablespoons
different-sized dried beans
(kidney, black, and lima do very well)
or paper clips or small safety pins

*T*ake a teaspoon and fill it with kidney beans. Don't scoop the beans. Instead drop them onto the spoon one at a time. You want to work carefully and hold your spoon level, because as soon as a bean falls out, you must stop. Now count how many beans you managed to get on the spoon and record the number on a piece of paper. The last time I did this, there were seventeen kidney beans on my spoon before one fell off.

Now give the spoon to your child and defy him to do better. Encourage him to work carefully and show him how to hold the spoon steady. Remind him that as soon as one bean falls off the spoon, he must stop filling. Eventually a bean will dribble off, and he must count how many beans stayed in the spoon.

How about a second contest? This time see who can fill a teaspoon fullest with little black beans.

Next, switch from teaspoon to tablespoon. The bean pile begins to climb. When the time comes for counting, your child will get practice counting high, high, higher. A tablespoon can hold over one hundred black beans. Imagine! I'm not sure how many grains of rice it can hold. You might want to find out, if you're more patient than I am.

If your child gets confused counting past twenty, thirty, or forty, offer as much help as he needs. You might provide the next number when he gets stuck, or say all the numbers and let your child chant along with you.

One memorable day, Linda, an eager first-grader, packed one hundred twenty-four black beans on a tablespoon. Linda insisted on counting this gigantic bean pile twice to make sure the number was exactly right. She was delighted. I was delighted. I didn't even mind picking beans off the floor for the next couple of days.

**D**oris didn't sit in her chair; she swayed. She didn't walk around the classroom; she bounced. Not surprisingly, Doris liked active games, and GRASSHOPPER was her favorite.

Before starting to play GRASSHOPPER, we needed empty floor space for jumping around. That meant pushing aside a table and a few chairs. We also needed a handful of index cards. Sixteen are enough to start a game. On each card I wrote one of the numbers from zero to fifteen. Then we taped the cards, hopping distance apart, in haphazard fashion across the floor.

**GRADES**
one and two

**MATERIALS**
index cards
pen
masking tape

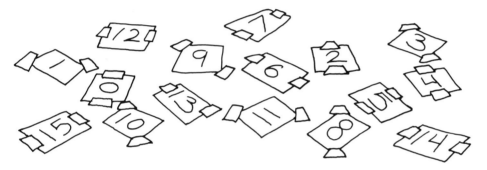

Doris liked helping with these preparations, but she liked it even more when I barked out a series of instructions:

"All right, Doris, be a grasshopper and hop onto the five. Good. Now, find the number that's one less than five. Go ahead, hop onto it. Skip to the number that's one more than twelve. Can you get to eight in one jump? Great! How about twirling to the number that's one less than fifteen? Now, count backwards from fourteen to seven and hop onto each number as you go."

Sometimes I let Doris bark the orders, and I hopped. In fact, the game worked wonderfully when we took turns ordering each other about. I told Doris to jump to three. She told me to jump to one less than eight. Back and forth we ordered and jumped until one of us, generally me, declared enough was enough. For the most part, I was an accurate hopper. But I promised Doris that I would make occasional bloopers. Once Doris told me to skip to one less than twelve, and I pounced on ten. Since Doris enjoyed catching my mis-hops, she was careful to calculate the correct answer to each of her commands.

After a while, Doris was pretty secure from zero through fifteen, so I made a new set of cards. This time I used twenty-one cards numbered five to twenty-five. Some days I put out cards from zero to twenty-five, other days, twenty-five to fifty. Once, I even made a set of high-numbered cards that had us hopping from one hundred to one hundred twenty.

This gymnastic workout gave Doris some good math experiences. Hopping around made for lots of practice in recognizing numbers. Young children can find it hard to remember that the word *twenty-one* goes with the figure *21*, while the word *twelve* goes with *12*. Grand numbers like 87 or 103 can also throw them for a loop. GRASSHOPPER gave Doris practice. The game also helped her relate each number to its neighbor. When I ordered Doris to hop to the number that is one more than eleven, she had to think about the number eleven, count one number past, and end up at twelve. At first this was difficult for her. She couldn't just say "eleven, twelve." Instead she counted from one to eleven before pouncing on twelve.

But the more we played GRASSHOPPER—the more she heard me count, and the more counting she did herself—the more sophisticated Doris became. Soon she was counting directly from eleven, not to mention from twelve or from one hundred.

If you have a hoppy child who can use help in recognizing numbers, and who needs practice starting to count from numbers other than one, then GRASSHOPPER is the game for you. Happy hopping!

*I*n counting, practice makes perfect. Luckily, most children love to count, so finding occasions to practice is easy. All you need to do is think of countable things. How many spoonfuls of cereal will you eat before the bowl is empty? How many forks are on the table? How many steps from the refrigerator to the TV? How many hops from the park bench to the swing? How many cans of cat food in the kitchen cabinet? How many toys on the floor? Try counting things as you put them away on shelves and in closets. This activity may seem a little obsessive, but it's useful.

Here's a way to make the usefulness even greater. Have your child estimate how many spoonfuls, steps, or toys there are before he counts. You can explain that an estimate is similar to a guess. When you're estimating, you look at the cereal bowl and try to figure out, without counting, how many spoonfuls are left to eat. You look at the distance to the refrigerator and, before walking, try to figure out how many steps from here to there.

Like guesses, estimates are sometimes right and sometimes wrong. Assure your child that it doesn't matter if his estimate is exactly right. The only thing that matters with an estimate is that he think about his answer first. Thinking first is what makes an estimate different from a guess. Instead of blurting out the words six, eight, ten, or twenty-three, estimating forces a child to give thoughtful consideration to the number of peas left on his plate, or marshmallows he can mash into a paper cup.

Some children hesitate to make estimates. They are overly concerned with being right and get discouraged when their estimates are wrong. One of the best ways to get kids going is to estimate yourself. Then your child can see that Daddy's estimates aren't perfect either. Try this over a bowl of popcorn. As you eat, make an announcement, "I think I can hold fifteen kernels of popcorn in my hand. How many kernels do you think I can hold?" After you and your child make popcorn-holding estimates, count and see how your estimates compare with the facts. "Look at that. I can hold twenty-nine popcorns. My estimate was way off."

It's a good idea to share your reason for making a particular estimate with your child. "Let's see, you took fifteen hops to get to the sofa. It's longer to the kitchen. I think you'll take thirty hops to get to the kitchen." Even if your child doesn't understand your reason, he'll understand that your thinking isn't arbitrary. That gives him a model to follow for his own thinking.

As children mature, their estimates take on a certain expertise. They begin to recognize that if it takes ten chews to finish one pretzel, it will take about twenty to eat two. If it takes five giant steps to get from the bed to the bathroom, it will take a lot more baby steps to get back. This kind of logical thought about numbers and sizes requires lots of sophistication. Sophistication develops slowly. What's the first step to gaining numerical savoir-faire? Try counting the number of pennies you can stack on your index finger.

# CLEANING COUNTS

*I* have discovered the trick to cleaning a messy classroom. When cleanup time rolls around, I shout: "This room gets cleaned by the time I count to fifty. Ready? Go! One . . . two . . . three . . . four . . ." Scurry, scurry, children begin to clean. By the time I hit fifty, the room is miraculously tidy.

Counting out loud has other effects as well. Listening to me, my students got daily practice with numbers. Frequently the children would count along with me, turning my solo act into a group performance. I began increasing the complexity of my recitations. I called out, "Today's cleanup is a count of one hundred. Only, today I'm counting by twos. Ready? Go! Two . . . four . . . six . . ."

I tried all sorts of skip-counting. I counted by tens, by threes, by fours. Sometimes I counted backwards. I counted backwards from fifty to zero. I counted backwards by five from one hundred to zero, and by threes from thirty to zero. I didn't expect children to count along when I engaged in these numerical acrobatics— although some tried. I just wanted to show that numbers can be counted in this, that, and every which way.

There was still another benefit. When the children listened to me skip-count, when they tried to imitate me, they weren't just counting, they were multiplying. Skip-counting to clean a classroom or a bedroom is a terrific way to introduce first- and second-graders to multiplication. When they reach third or fourth grade and confront the awesome task of memorizing multiplication tables, CLEANING COUNTS can come to the rescue. You can help your child with the pesky fours table by counting four . . . eight . . . twelve . . . sixteen . . . while your child picks up his toys. Or recite the sixes table while your child clears the kitchen table.

Incidentally, you can use the same method to get a child to put on his gloves or finish his peas. Really, counting out loud is one of the most useful techniques a parent or teacher can employ. It instructs; it amuses; it avoids fights; it gets children to perform their onerous tasks. Life would be easier if there were a lot of techniques as practical as clearing your throat and saying: "Ready? Go! One . . . two . . . three . . ."

# FIND A PENNY

GRADES

one, two, and three

MATERIALS

paper cups
pen
penny

**H**ere are ten upside-down paper cups placed in a row, each cup labeled one through ten.

A penny is hiding under one of the cups. Can you guess which one? If your guess is wrong, I'll give you a hint. If you can find the penny in four guesses, you win. Go ahead, guess.

You think the penny is under the third cup? No, it's further away than the third cup. That's your hint: the penny is further away than the third cup. Take another guess.

You think the penny is under the eighth cup? No, it's closer than the eighth cup. That means the penny is further than the third cup but closer than the eighth.

You think the penny is under the sixth cup? No, it's closer than the sixth cup. You have one more guess.

You think it's under the fourth cup? Lift the cup. What do you see? Bravo, a copper Abe Lincoln. Now, I'll close my eyes and you can hide the penny. Then I'll guess and you'll give me hints.

This game is FIND A PENNY, and it has a child-approval rate approaching one hundred percent.

What do children learn from this paper-cup shell game? They learn ordinal numbers. Everyday life doesn't present many opportunities to learn first, second, third, fourth. Children are usually a lot more familiar with one, two, three, four. But with FIND A PENNY they practice ordinal numbers for the fun of it.

For many children, the clues in FIND A PENNY are hard to figure out. Very young children, younger than five, can't simultaneously think about closer and further. These children, therefore, can't coordinate clues like closer than the sixth cup but further than the third cup. So they guess—is it the eighth cup? And they hope for the best. Even older children—first-, second-, and third-graders—get confused trying to sort out FIND A PENNY clues. If your child gets confused, there's no point berating him. He'll just need a little extra help while you play.

The first time I played with Jennifer I realized that, unless I helped, she wouldn't FIND A PENNY except by luck.

**Peggy:** OK, Jennifer, the penny is hidden now, so you can open your eyes and take your first guess. Where could that penny be hiding?

**Jennifer:** Is it under the third cup?

**Peggy:** No, it's not. The penny is further away than the third cup. Can you point to the cups that are further away than the third? (Jennifer pointed.) Great. The penny might be in any one of those cups, but it's not in the third cup, or in the second or in the first. Let's turn those cups over so you don't forget.

**Jennifer:** Is it under the seventh cup?

**Peggy:** No, it's not. The penny is closer than the seventh cup. Can you show me the cups closer than the seventh? (Jennifer pointed to the correct ones.) Good. Now let's turn over all the other cups and see what we have.

With a delighted giggle, Jennifer turned over the seventh, eighth, ninth, and tenth cups.

**Peggy:** Now what's your guess?

**Jennifer:** Is it under the fifth cup?

**Peggy:** No, it's closer than the fifth cup. That means you can turn over the fifth cup. What other cup can you turn over?

**Jennifer:** Can I turn over the sixth cup?

**Peggy:** Exactly.

**Jennifer:** I know where the penny is! I know where the penny is! It's under number four.

Jennifer and I played many games with turned-over cups. Even when she hid the penny and I hunted, we used the turn-over-the-cup method. Eventually Jennifer outgrew her dependence on seeing the empty cups. She developed the ability to visualize, in her mind's eye, which cups might be hiding the penny.

If your child has trouble with the game, even with the aid of turned-over cups, it probably means he's a bit young for FIND A PENNY. Try it again in six months: you may be surprised how easy FIND A PENNY will be then.

**S**ECRET NUMBER is an advanced, penniless, cupless, version of FIND A PENNY. Here's how it works.

**Peggy:** I'm thinking of a secret number between one and ten. You have to guess my number. You only have four guesses. Every time you guess, I'll give you a hint.

**Mari:** Is your number nine?

**Peggy:** No, my number is smaller than nine. That's your first hint. You now know that my number is smaller than nine. Go ahead and take a second guess.

**Mari:** Is your number four?

**Peggy:** No, my number is bigger than four. Now you have two hints. You know my number is smaller than nine and bigger than four. Time for your next guess.

**Mari:** Let me think. Is your number six?

**Peggy:** No, my number is bigger than six. Now that means you just have one more guess. Good luck.

**Mari:** I know it's either seven or eight. I don't know which to ask! Can't I have two guesses? Please?

**Peggy:** No way, one more guess is all you get.

**Mari:** OK. Is it seven? I hope, I hope, I hope.

**Peggy:** Sorry, Mari, my number is eight. You came *so* close. Now you can think of a secret number for me to guess. Tell me when you're ready, and I'll start guessing.

This game is a lot harder than FIND A PENNY because the child must hold all the numbers in his head. If your child isn't ready to hold numbers in his head, you'll have to wait a few months and try the game again.

The child's ability to hold numbers in his head is only the first step in mastering SECRET NUMBER. The second step comes when he discovers the about-halfway-there strategy. What is this strategy? You begin by asking if the secret number is five, because five is about halfway between one and ten. You find out it's bigger, so you ask if the number is seven because seven is roughly halfway between five and ten. You find out it's bigger, and so you ask if the number is nine because nine is roughly halfway between seven and ten. You find out it's smaller, and you know positively the number is eight. The about-halfway-there strategy ensures that you will figure out the number in four guesses.

You don't need to give up on the game, however, when the numbers one to ten lose their mystery. Instead, extend the numerical terrain from one to twenty, or one to fifty, or one to one hundred. Of course with extra numbers come extra guesses. I usually allow six guesses with numbers from one to twenty, seven guesses with numbers from one to fifty, and eight guesses with numbers from one to one hundred.

If your child gets a kick out of SECRET NUMBER, I think you'll find it handy during car trips, standing in movie lines, or waiting for a shoe salesman.

# chapter 5.

# Adding and Subtracting

"Something's wrong. My child is halfway through second grade and he still uses his fingers to add and subtract."

"My child can add and subtract without her fingers, but she doesn't understand how to use arithmetic. The other day I showed her five dollars and asked how much more money we needed to buy a seven-dollar toy. She answered twelve dollars."

"My child's teacher says he's doing all right in arithmetic. But he hates math. I try to help him, but he complains so much it drives me crazy."

These are common worries for parents of early graders. The worries touch on what I call, with a little orthographical stretching, the three -tion problems: memorization, comprehension, and appreciation. When your child studies math, you want him to do well in all three -tions. You want him to memorize basic number combinations. You want him to comprehend how to use addition and subtraction correctly. And you want him to appreciate—to enjoy—the study of mathematics.

It would be wonderful if memorization, comprehension, and appreciation developed naturally in children. But this rarely happens. Most children need a little help with one or another, if not all, of the three -tions. Thus the games in this chapter.

Four of the games—MAKE TEN, TARGET GAME, NUMBER STORIES, and WHAT DID I DO?—are comprehension games. They help children understand how addition and subtraction work. The rest of the games are memorization games. They help children remember the addition and subtraction equations that everyone must sooner or later know. Let's face it, that's drill work. But drills in game form can turn pain into pleasure. Of course playing memorization games can help your child with comprehension, just as comprehension games can help your child with memorization.

Every game you play should also be an appreciation game. If you or your child don't care for a particular game, don't play it. Just because I think a game is fun doesn't mean you'll agree.

Before children memorize addition and subtraction equations, they usually calculate answers on their fingers. That's fine. You can also encourage your child to use pennies, dried beans, or paper clips instead of fingers. The main thing is to give your child lots of practice. Lots of practice will eventually lead to fewer dried beans. Don't rush your child, however. With patience on your part, and positive feelings about math on your child's part, all good things will come.

The games in this chapter are geared for a variety of grade levels, which I've indicated at the start of each game. But don't take these designations too seriously. Sometimes your third-grader will benefit from playing a first-grade game. Children are never too old for a good game. Sometimes they're too young, though. You should be careful about trying a third-grade game with a first-grader. When you push a child, you may gain a little memorization, and a little comprehension, but you will surely lose appreciation, which will cost the child in the end.

# MAKE TEN

**M**AKE TEN is a quick, easy game that helps first-graders understand addition and subtraction. That's why I wanted to play it with Blanche, a little girl who needed a little help. I got out a playing die, gathered up twenty paperclips from a bowl on my desk, and drew two game boards on blank sheets of paper.

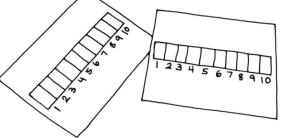

**GRADE**

one

**MATERIALS**

paper

pen

one playing die

at least twenty counters

(poker chips or dried beans will do)

"I want to go first," Blanche said.

"OK," I said. "Throw the die."

Blanche threw a four. She took four paperclips and placed them in the first four boxes of her playing board.

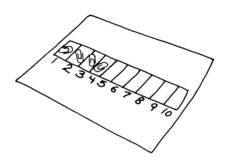

"That was a lucky throw," I said. "Can you tell me how many more clips you need to fill up your board?".

Blanche counted, and when she was done she said, "I need six more. When I get six more, I'll win!"

"Only if you make ten before I do. Give me the die so I can have a turn." I threw the die and got a two. I took two clips and placed them on my board. "Looks like you're winning so far. You have one . . . *two* more boxes filled than I," I said as I pointed to the empty boxes on my board that corresponded to filled boxes on Blanche's board.

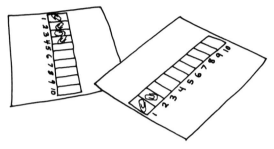

"My turn now," said Blanche. She rolled the die and got three. She took three clips and added them to her board.

"You used to have four, and now you have seven. Your board is filling up very fast. I need five clips to catch up to you. If I get a six, I'll have more than you." I threw and got that six. I took clips and added them to my board. "I have eight now. I only need two more to win!"

"Yeah, but I only need three more," replied Blanche. "And it's my turn." Blanche rolled the die and threw a five. "I win! I win! I filled up my board!"

"Congratulations! Now let's try the take-away version of MAKE TEN. Do you remember how to play that way?"

Blanche nodded. In take-away MAKE TEN you begin the game with a board full of paperclips.

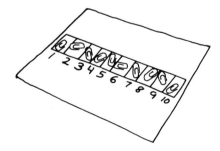

"Can I go first this time?" I asked.

"Well . . . OK," grumbled Blanche.

I threw the die and got a five. "Oh, that's lucky. I get to take five clips away. I just have five left."

"You're going to win," she complained.

"Maybe I will. Just like you won last time. But there's no way to know for sure until you throw the die."

Blanche threw a six. A smile returned to her face as she removed six clips from her board.

"You only have four clips left and I have five. Still, if I throw a five or a six, I'll win. I hope I get lucky." I threw a three. "Here are five clips, and I take away one, two, three clips. That leaves me with just two clips."

Blanche gave the die a vigorous shake, and got another six.

"I won! I won again! I can take away all my clips!"

"Lucky you. You're the MAKE TEN winner in both versions." And if having a little practice in adding and subtracting was a good thing, she won that way, too.

# WHAT DID I DO?

**GRADES**

one and two

**MATERIALS**

a small collection of objects
for counting (paper clips, pennies,
dried beans)

**L**ook at the paperclips in my hand. How many are there?

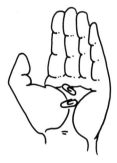

Close your eyes so I can make a change. All right, you can open.

Did I add clips or take clips away? How many clips did I add? Let's try once more. Look at my hand. How many clips?

Close your eyes so I can fiddle a bit. OK, open up.

What did I do? You're right—I took away five clips.

You just played WHAT DID I DO? The first time you play this game with your child, you may be surprised at how much effort it takes for the child to figure out what you did. There are two ways you can help. First, give the child time to think. The less hurry, the less worry. Second, have your child re-enact what you did.

Here's how a re-enactment works. Let's say you start with four clips in your hand. While your child closes his eyes, you add three more. When the child opens his eyes, he's confused and can't figure out what you did.

"How many clips did I have in my hand before?" you ask.

"Four," your child answers.

"Take four clips and put them in your hand. Now hold your hand next to mine.

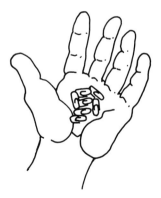

Do you think I added clips or took clips away?"

"You added."

"Right. Go ahead and add clips to your hand until you have the same number of clips as I. Count each clip you put in your hand, and you'll know what I did."

Re-enactment works when you remove clips, too. Let's say you begin with six clips and, while your child's eyes are shut, you remove five. Your child opens his eyes, looks at your hand, but can't figure out what you did. Here's how to help your child out.

"Do you remember how many clips I used to have?" you ask.

"Six," the child answers.

"Well then, put six clips in your hand."

The child picks up six clips.

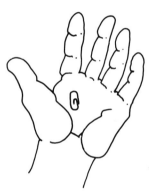 

"Now take away clips until we both have the same amount. Keep track of how many clips you remove. If you count carefully, you'll know just what I did."

WHAT DID I DO? is a fine game to play while waiting for food in a restaurant. All you need to do is pull some pennies out of your pocket and you'll liven up a boring time. You might also play in the car, if you're not the driver. It'll make a road trip more fun for everyone.

Why do most children have an easier time with addition than subtraction? There are many complicated and abstruse reasons having to do with the number system. There's also an uncomplicated reason: greed. The concept of "more" sits very well with children. The concept of "less" is not so appealing. Addition therefore enjoys a psychological advantage. Children naturally give themselves lots of practice with it. They want one more glass of juice, two more crackers, four more crayons. They do not give themselves such practice with subtraction.

The purpose of TARGET GAME is, then, to tie the spirit of greed to the practice of subtraction. You and your child begin with ten paperclips each. The object is to rid yourself of these clips.

You'll need to make a board that looks like this:

Draw the board on paper, oaktag, or poster board, and use colored markers to design the target and write the numbers. You'll also need twenty paperclips and two pennies.

Here are the rules: you and your child take turns throwing pennies at the target. The number you land on tells how many clips you lose. Land on five, lose five clips. Land on two, lose two clips. Land on zero, lose nothing. Land on a line or go off the board, you

**GRADES**
one and two

**MATERIALS**
paper, oaktag, or poster board
colored markers
two pennies
twenty paperclips, dried beans, or
bits of colored paper

throw again. Land on ten, you lose the whole batch and win the game. Losing, in this game, is winning.

I keep up a running commentary while playing this game: "I had eight clips and landed on two. That means I can get rid of two clips. Now I have six. How many clips do you have? You have seven. Lucky you, you threw a four. You can subtract four and you only have three left. You have less than I. I hope I get ten. If I get ten, I can get rid of all my clips. Oh no—I got zero."

What does all this talk accomplish? The idea is to focus attention on the diminishing scores. My talking helps children get acquainted with the subtraction process. I don't expect children to remember that six take away four is two, or that seven take away four is three. These are facts for memorization. TARGET GAME is about something else: getting used to subtraction.

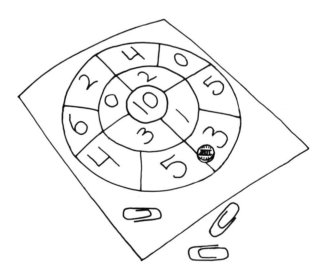

**G**inny was in first grade and already an exceptional math student. She delighted in counting, adding, subtracting. She loved to compute numbers. Her mastery and enthusiasm were unusual and I wondered what drew her so eagerly to the subject. Part of the answer came when I overheard a chat between Ginny and her father in the school stairwell.

**Dad:** Once there was a giant who was taller than the tallest tree in the world. Naturally this giant had a huge appetite. Every day for breakfast he ate three bowls of corn flakes, four bowls of grits, and two bowls of oatmeal. How much cereal did the giant eat?

**Ginny** (busily counting on her fingers as she talked): I know, I know. Wait, wait, I can figure it out. Nine! Nine! He ate nine bowls!

**Dad:** Right you are. This giant—his name was Gulumph—loved oatmeal best of all. No sooner had Gulumph finished washing the breakfast dishes than he was hungry again, and decided to prepare a snack of oatmeal and raisins. He combined three cups of oatmeal and two cups of raisins in a great big bowl. How many cups of oatmeal-raisin snack were in the bowl?

**Ginny:** That's easy, there were five cups.

**Dad:** That's right. But before he could eat any snack, his cat Meowapuss climbed on the table and ate two cups of cereal. Gulumph shooed Meowapuss away, and ate the rest himself. How much did Gulumph get to eat?

**Ginny:** Let me think. How much was in the bowl? I forget.

**Dad:** There were five cups in the bowl and Meowapuss ate two cups.

**Ginny:** OK, I get it. (She fiddled with her fingers a minute.) Gulumph got three cups.

**Dad:** Exactly. Well, Gulumph finished eating, but he was angry at Meowapuss. He made the cat wash all the plates in the house. But Meowapuss was bad at dishwashing. She broke three plates. So now Gulumph only had five plates left. How many did Gulumph have before Meowapuss washed the dishes?

**Ginny:** I don't get it.

**Dad:** Let me help. Meowapuss broke three plates, right? (Ginny nodded her head.) And there were five plates left, isn't that so? (Ginny nodded again.) How many plates did there used to be altogether, then?

**Ginny:** Well, he used to have three broken ones and five good ones—so he used to have eight. Is that it?

**Dad:** Bravo!

By now Ginny and her dad had descended the three flights of stairs. I decided to pull Ginny's dad aside and ask him about this fanciful tale of Gulumph and Meowapuss. He told these stories all the time, he said. He made them up as he and Ginny walked to school or went to the supermarket, or waited for food in restaurants. Sometimes Ginny came up with her own stories, too. The warmth and humor Ginny and her father shared during this mathematical storytelling went a long way toward explaining Ginny's enjoyment of numbers.

I wondered how a less mathematically inclined child would react to such stories. I conducted an experiment on a second-grader named Adam, who struggled with numbers and complained about having to study them. I started telling Adam a story about bank robbers. It was amazing—Adam was actually enthusiastic when it came time to figure out how many bags of money the robbers took from the vault and how many guards chased the robbers down the street. So here was a technique, NUMBER STORIES, that worked with both a mathophile like Ginny and a mathophobe like Adam.

You'll want to keep a few points in mind if you decide to tell your own NUMBER STORIES. Storytelling time should be relaxed and playful. If your child begins to feel NUMBER STORIES are a way of testing him, he won't enjoy himself, and the value of the stories will be lost. The best thing is to keep most of the questions easy. The numbers should be small, the problems straightforward. Here's a straightforward problem: three car-thieves and two jewel-thieves got together to rob a bank. How many thieves were there? Here's a more complicated problem: the thieves stole ten bags of gold altogether. The jewel-thieves stole seven bags. How many did the car-thieves get? If you keep most of your problems straightforward, your child won't feel anxious when it comes to solving an occasional complicated problem.

The stories you tell can be realistic or fantastic, rooted in events of your day, or in the daily events of the witch next door. Your stories don't have to be especially brilliant. I've found that even the worst plots hold children's attention—at least for a few minutes. Here are a few stories you might try on your child before inventing ones of your own.

★

Once there was a witch who lived in an old castle with three black cats and four ghosts and one goblin. How many creatures lived with the witch? This witch loved casting spells on people. On Wednesday, she called her three cats to travel with her, and set out on her broom to cast spells.

Soon she saw a boy with a bag full of candy. The boy had two chocolate kisses, two caramels, and two gum drops. How many candies did the boy have? The witch decided to cast a spell on the candy. She turned four of the candies into spiders. The rest she left alone. How many candies does the boy have now? The spiders started to crawl out of the bag. This scared the boy out of his wits. But the witch forgot one thing. Her cats were terrified of spiders. In fact, the cats were so scared, they jumped off the broom. This made the broom wobble and jerk. Suddenly, the witch tumbled off the broom. She got five bumps on her head and six on her nose. How many bumps did she get altogether? Poor witch.

★

I went shopping yesterday. First, I went to the shoe store. I looked at lots of shoes. I tried on two pairs of red shoes and three pairs of blue shoes. How many pairs did I try on? I decided not to buy any shoes. Instead, I decided to buy a dress. So I went to the dress store. There were ten dresses on the rack. Six dresses were really ugly. The others were nice. How many were nice? I decided to buy one of the dresses. The dress cost forty dollars. I gave the salesperson fifty dollars. Did I give her too much or too little? I spent ten minutes in the shoe store and ten minutes in the dress store. How much time did I spend shopping?

★

A man went to the zoo. He saw five zebras and six monkeys. How many animals did he see? Suddenly the zebras' cage opened up. Two of the zebras ran away. How many zebras were left in the cage? The man started to chase the runaway zebras. In his chase he ran into a popcorn man. The popcorn man fell down and so did his popcorn. Five bags of popcorn fell into the street and three bags fell on the sidewalk. How many bags fell down? When the zebras saw the popcorn, they came running back. They wanted to eat popcorn. Before anyone could stop them they ate six bags. How many bags were left? Then the man had an idea. He made a trail of popcorn from the popcorn man to the cage. The zebras kept eating the popcorn. Before they knew it, they were back inside their cage.

★

What do fantasies like these stories have to do with math? I quote Einstein: "When I examine myself and my methods of thought, I come to the conclusion that the gift of fantasy has meant more to me than my talent for absorbing positive knowledge."

*L*ynn was a second-grader doing well enough in math, except that she found the subject boring. She needed to memorize a lot of numbers, and this struck her as impossibly tedious. She did her work; she learned; but the delight of learning always eluded her. NUMBER LADDER changed all that. It drilled her on number facts, but it also kept her interested.

I drew a board like this:

**G R A D E S**

one, two, and three

**M A T E R I A L S**

paper
pen
one playing die

The numbers (one through ten) were in no particular order. Then I gave Lynn a single die and told her to throw. She tossed a three. That meant she had to add three to each number on the ladder. Starting on the bottom rung, she added three plus two. Second rung: three plus six. Third rung: three plus one. She added these numbers, and as long as she did it correctly, she continued to climb. If she added all rungs correctly, she made it to the top of the ladder. If she faltered in her addition at any rung along the way, however, she tumbled down the ladder—*kerplunk.*

Lynn and I took turns climbing the ladder. Of course, I knew all the addition. I didn't even use my fingers. Consequently, a small adjustment in the game was called for. I promised that during my turn I would make at least one addition mistake, maybe more. If Lynn caught my mistake, I'd fall down the ladder. If Lynn let my mistake slip by, I'd get to the top. I always waited at each rung until Lynn gave me the OK to proceed. That way, she had plenty of time to catch my mistake. She just had to be as careful in her addition as I was careless in mine. Eventually I couldn't trick her, which thrilled her no end—and her teacher too.

# FAST
# TRACK

**GRADES**

one, two, and three

**MATERIALS**

paper, oaktag, or poster board,
approximately 12" × 20"

colored markers

index cards

scissors

a game token for each player
(a paperclip, a small model car,
a thimble)

**F**AST TRACK is one of those boom-or-bust board games full of lucky and unlucky moves. It's an exciting game, and it gives children plenty of practice in addition and subtraction.

You need to make game cards, which I do by cutting ten index cards into fourths. On each card write an incomplete arithmetic problem from the lists on pages 75 and 76. Most of these problems are missing the last number: 3 + 3 = [   ] or 6 − 2 = [   ]. Others are missing a middle number: 3 + [   ] = 8 or 6 − [   ] = 2. Problems missing the middle number are hard to solve, and FAST TRACK won't make them any easier. But the game makes them fun. The missing number in each problem is how far a player travels clockwise around the FAST TRACK board.

Creating the board takes some time. Oaktag makes the best board because it doesn't fall apart, but any large piece of paper will do. The board looks like this:

There are three *Start* boxes. A unique feature of FAST TRACK is that each player can pick his own *Start* box. There are three *Win!* boxes. The first player to land on one of these boxes wins the game. The rest of the boxes are, for the most part, standard board-game fare: *double move, lose a turn, go ahead 2 spaces, go back 3 spaces, move to Start, double move,* and *draw again.* There are also boxes called *stay put.* If you land on one of them, you don't move ahead until your next turn. Your board needn't be an exact copy of mine. You might have a few more boxes or a few less, more *go ahead*s and less *move to Start*s—that's perfectly okay.

When I make a FAST TRACK board, I color-code it. I write *Win!* in red, *Start* in brown, *lose a turn* in blue, *go ahead 2 spaces* in purple, *go back 3 spaces* in yellow, *move to Start* in green, *double move* in orange, *stay put* in pink, and *draw again* in black. The color-coding helps even beginning readers move around the board more or less independently.

Your child might enjoy helping you make the game board. I made the one drawn here with a second-grader named Mary. Working on the board made her feel useful. She liked using the colored markers. But what she liked best was playing the game once we finished the board.

Here's how it went. We picked playing tokens. Mary used a tiny cat-shaped eraser. I used a thimble. Next we picked *Start* places. Mary chose the *Start* at the top of the track. I put my token on the *Start* at the bottom left-hand side of the track. Mary wanted to go first, which was OK with me. She picked the top card from the shuffled card deck. The card read $4 + 6 = [\ \ ]$. She computed, she said the right answer, and she moved her eraser clockwise ten boxes. She landed on another *Start.* My turn came next, and I drew $10 - 8 = [\ \ ]$. Moving two spaces clockwise, I landed on *go back 3 spaces.* I went back three boxes and landed on *draw again.* This time I picked $2 + 6 = [\ \ ]$. Moving forward eight boxes, I found myself on *lose a turn.* Bad luck! Mary drew her second card. It was $7 - [\ \ ] = 5$. She was taken aback by this equation. She needed help to solve it, and I was happy to assist.

"What number starts the problem?" I asked.

"Seven," she answered.

"Good," I said as I held up seven fingers. "What number do you want to end up with?"

"Five," was her reply.

"Right. How many fingers do you need to push away to end up with five?"

"Two," she squealed, grabbing her playing token and advancing two boxes. She landed on *go ahead 2 spaces.* Going ahead two spaces, she landed on *double move.* This meant Mary could move ahead another two spaces since her last card was 7 — [ ] = 5. Now she landed on *move to start.* Mary moved ahead again, this time to the next *Start* box.

Mary's turn was over, but since I had to lose one turn, she got to go again. She drew 4 + [ ] = 6. Mary figured out this answer on her own, and moved ahead two spaces. She landed on *go back 3 spaces.* After traveling backwards, she landed on *draw again.* She drew 7 — 6 = [ ]. She moved up one space, and returned to the same *Start* box she'd left just a few moments ago.

At long last it was my turn. I drew 3 + [ ] = 6. Instead of stating the answer, I worked the problem out on my fingers.

"Here are three fingers," I said raising my thumb, index, and middle fingers in the air. "If I add one finger, I don't get six. If I add two fingers, I don't get six. If I add three fingers, that's six." Now I was ready to move forward three boxes. I landed on *double move,* and went ahead another three boxes. I ended up on *go back 3 spaces.* This sent me right back to *double move.* Since my last card was 3 + [ ] = 6, *double move* meant going ahead three boxes once more, which meant landing on *go back 3 spaces* again. Back and forth, back and forth, forever. Clearly, we needed a special rule to resolve this seesaw move. So I proclaimed that when caught in such a back-and-forth trap, a player stops at the furthest ahead box and stays put. In my case, this meant stopping on *go back 3 spaces.*

My turn was over, so Mary drew a card. She got 7 − 4 = [ ]. It was easy for her to figure out the answer. Moving ahead, she landed on *double move.* She repeated her three-step move and landed on *Win!*

FAST TRACK was a big hit with Mary. If you think it will go over big in your house, why not get out your index cards, poster board, colored markers and start making a game? Below are equations for three different FAST TRACK games. You should begin with the Easy List. When your child seems ready for tougher equations, go on to the Less Easy List. Finally, when your child is an addition and subtraction pro, make a deck using the Fiendishly Difficult List. Don't rush the child, though. It's better for a child to review and practice easier equations than to struggle painfully and joylessly with problems that are too difficult.

## Easy List

| | | |
|---|---|---|
| 8 + 0 = [ ] | 5 + 5 = [ ] | 8 − 2 = [ ] |
| 0 + 7 = [ ] | 6 + 3 = [ ] | 4 + [ ] = 5 |
| 1 + 3 = [ ] | 6 − 3 = [ ] | 6 + [ ] = 10 |
| 2 + 5 = [ ] | 7 − 2 = [ ] | 5 − [ ] = 3 |
| 2 + 3 = [ ] | 8 − 3 = [ ] | 6 − [ ] = 3 |
| 2 + 4 = [ ] | 9 − 2 = [ ] | 2 − [ ] = 2 |
| 6 + 2 = [ ] | 7 − 7 = [ ] | 4 − [ ] = 0 |
| 2 + 7 = [ ] | 3 − 1 = [ ] | 5 + [ ] = 7 |
| 2 + 8 = [ ] | 6 − 2 = [ ] | 3 − [ ] = 2 |
| 3 + 3 = [ ] | 3 − 2 = [ ] | 1 + [ ] = 4 |
| 3 + 4 = [ ] | 10 − 2 = [ ] | 3 + [ ] = 5 |
| 9 + 1 = [ ] | 5 − 0 = [ ] | 8 + [ ] = 9 |
| 3 + 5 = [ ] | 9 − 0 = [ ] | 4 − [ ] = 2 |
| | 5 − 2 = [ ] | |

## Less Easy List

2 + 10 = [   ]        10 + 4 = [   ]        12 − 4 = [   ]
3 + 10 = [   ]        10 + 6 = [   ]        9 − 7 = [   ]
4 + 10 = [   ]        10 + 8 = [   ]        13 − 3 = [   ]
5 + 6 = [   ]        10 − 3 = [   ]        7 − [   ] = 4
4 + 5 = [   ]        12 − 2 = [   ]        6 + [   ] = 10
6 + 4 = [   ]        13 − 3 = [   ]        8 − [   ] = 5
7 + 4 = [   ]        10 − 4 = [   ]        4 + [   ] = 9
4 + 8 = [   ]        10 − 6 = [   ]        10 − [   ] = 7
3 + 8 = [   ]        8 − 3 = [   ]        10 − [   ] = 4
5 + 4 = [   ]        9 − 4 = [   ]        9 − [   ] = 9
5 + 7 = [   ]        9 − 5 = [   ]        2 + [   ] = 8
9 + 9 = [   ]        9 − 3 = [   ]        7 − [   ] = 4
4 + 9 = [   ]        8 − 7 = [   ]        5 + [   ] = 10
                        6 − 5 = [   ]

## Fiendishly Difficult List

6 + 7 = [   ]        7 + 10 = [   ]        18 − 10 = [   ]
6 + 8 = [   ]        8 + 10 = [   ]        19 − 10 = [   ]
5 + 7 = [   ]        13 − 5 = [   ]        16 − 10 = [   ]
5 + 8 = [   ]        17 − 7 = [   ]        6 + [   ] = 14
8 + 9 = [   ]        12 − 7 = [   ]        12 − [   ] = 10
6 + 4 = [   ]        14 − 5 = [   ]        12 − [   ] = 8
9 + 7 = [   ]        14 − 9 = [   ]        5 + [   ] = 11
5 + 9 = [   ]        13 − 6 = [   ]        8 + [   ] = 8
8 + 7 = [   ]        10 − 5 = [   ]        13 − [   ] = 10
8 + 8 = [   ]        11 − 3 = [   ]        4 + [   ] = 14
3 + 9 = [   ]        14 − 8 = [   ]        2 + [   ] = 12
6 + 9 = [   ]        15 − 9 = [   ]        8 − [   ] = 1
5 + 4 = [   ]        16 − 9 = [   ]        6 + [   ] = 10
                        12 − 8 = [   ]

**D**OUBLE IT is a board game with a board that looks like a snail:

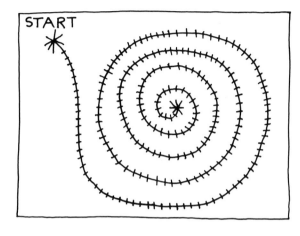

**GRADES**
one, two, and three

**MATERIALS**
paper, oaktag, or poster board
index cards in two different colors
pen
two game tokens (a paperclip and
a thimble, for instance)
optional: something to use for
counting, such as dried beans or
bingo chips

You can draw this on a big sheet of paper, or else use oaktag or poster board for something nicer and more permanent. Next, you need to make a deck of forty-four Number Cards marked zero through ten—that's four cards for each number. I make this deck using yellow index cards cut into quarters.

You also need a set of twenty-four Instruction Cards that say *add, subtract, double one number*, or *lose one turn*. I cut six white index cards into quarters to make this deck. You'll want eight *add*s, six *subtract*s, seven *double one number*s and three *lose one turn*s.

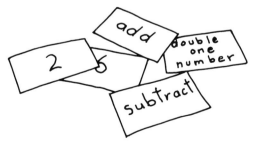

So, now: with your token at the starting position, pick two Number Cards and one Instruction Card. The Instruction Card might say *add*, in which case you add the two Number Cards and move your token that many notches. The Instruction Card might say *subtract*. In that case subtract the smaller number card from the larger, and move your token that far. If the instruction is *double one number*, take the bigger Number Card, double it, and move that far. If the instruction card says *lose one turn*, you stay put. Everybody takes turns in this game until someone completes the spiral and wins. You may have to reshuffle the cards before the game is over.

The short-term goal of DOUBLE IT is to get to the end of the spiral. The long-term goal is to help your child master addition and subtraction. When you play the game, your child may need props to add and subtract. That's OK. Let him use his fingers, your fingers, dried lima beans, a penny collection, or any other counting tools you have around the house. The more he plays, the more practice he'll get with numbers. Slowly, perhaps without his being aware of it, he'll begin to memorize the problems. In time, he won't need fingers or beans to add three and five. He'll find doubles like $4 + 4$, $8 + 8$, and $6 + 6$ easier and easier to remember. He may even begin using doubles to solve other problems—if three plus three is six, then three plus four must be seven. This thoughtful analysis of addition and subtraction problems results in the best and most permanent kind of learning.

Simply playing DOUBLE IT is no guarantee that your child will memorize all one hundred addition and fifty subtraction equations that might crop up in the game. But DOUBLE IT adds fun and subtracts boredom, adds thoughtfulness and subtracts mindless drill.

ax was on the verge of rebellion. For months he'd been struggling to memorize addition and subtraction equations. He'd filled workbook pages. He'd studied flash cards. He'd practiced and drilled, but enough was enough. It was time for a little fun; yet it was still time, unfortunately, for drill. Thus MATH CHECKERS.

You play MATH CHECKERS on a regular checkerboard, only first you have to tape little bits of paper to all the black squares. On each bit of paper goes an addition or subtraction problem.

GRADES
one, two, and three

MATERIALS
a checkerboard
checkers
paper
scissors
pen
transparent tape

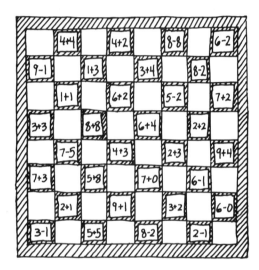

Play proceeds in MATH CHECKERS exactly as in traditional checkers, with one exception. In MATH CHECKERS, before a player can move his piece to a new square, he has to solve the addition or subtraction problem on that square. If a player makes a computational error, he still gets to move his piece—after his arithmetic is gently corrected.

On the next page is a checkerboard of a game-in-progress. In this game Max was black and I was red. It's Max's turn, and he's well positioned to jump my piece. He adds five plus eight on his fingers, and tells me it equals fourteen. I say no, five plus eight

*Adding and Subtracting*

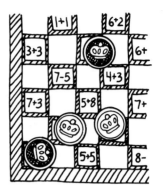

equals thirteen, and Max makes his move. It's my turn. I tell Max the answers to seven minus five *and* six plus two. Then I double jump. Great game, checkers.

Thanks to MATH CHECKERS, Max developed some enthusiasm for addition and subtraction. He liked MATH CHECKERS so much, I made it a regular feature of our weekly math sessions. From time to time, I untaped problems Max had already memorized and replaced them with new and harder equations.

When you play MATH CHECKERS, begin with easy problems, and slowly work up to harder ones. Below are four different checkerboards, ranging in difficulty from Easy to Fiendishly Difficult. You'll notice some problems show up on more than one board. Also, there are a few easy problems even on the hardest board. That's important. Games where every problem is a challenge cause stress, not progress. You can copy the problems on these boards to make your own boards for MATH CHECKERS.

Here's an idea that may help your child think more analytically about numbers. When it's your turn and you're solving MATH CHECKERS problems, do some thinking out loud: "Let's see, five plus five is ten, so five plus six must be eleven." Or, "I know that ten plus four is fourteen, so that means nine plus four must be thirteen." Your child may not follow your logic, but he'll see that *you* think through problems and may try thinking about numbers, too, just to be like Mom or Dad.

## Easy Board

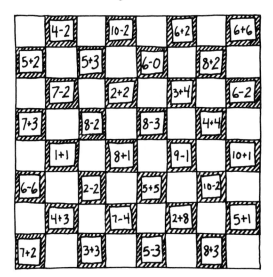

4-2  10-2  6+2  6+6
5+2  5+3  6-0  8+2
7-2  2+2  3+4  6-2
7+3  8-2  8-3  4+4
1+1  8+1  9-1  10+1
6-6  2-2  5+5  10-2
4+3  7-4  2+8  5+1
7+2  3+3  5-3  8+3

## Less Easy Board

10+5  13-3  10+2  7+7
7+3  16-6  8-3  8-5
7-4  10+10  9+3  8+3
12-2  9+9  6+4  4+5
8-8  10+4  7-3  5+5
8+2  10-2  8+8  7-2
6-2  6+5  10-6  7-5
10+3  10-8  6-4  10-4

## Getting Difficult Board

18-8  8-6  8+9  10-2
9+4  10-7  7+8  4+5
10-4  10+8  9+5  10+3
8+4  10+9  16-6  19-9
13-10  19-10  7+5  17-10
3+4  8+5  9-5  10+4
7+4  7-5  9-4  5+6
14-10  9-6  6+7  8-4

## Fiendishly Difficult Board

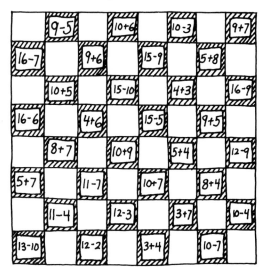

9-5  10+6  10-3  9+7
16-7  9+6  15-9  5+8
10+5  15-10  4+3  16-9
16-6  4+6  15-5  9+5
8+7  10+9  5+4  12-9
5+7  11-7  10+7  8+4
11-4  12-3  3+7  10-4
13-10  12-2  3+4  10-7

*Adding and Subtracting*

81

# PYRAMID

**GRADES**

two and three

**MATERIALS**

a deck of cards
with picture cards and
Jokers removed

pen

Iplayed PYRAMID with Sean one Monday, and Tuesday he paid the game a compliment.

"You know, I taught my grandma PYRAMID last night," he said. "We played and I won. It was fun."

What better recommendation could a game get?

Grandma may have been familiar with PYRAMID, since it's a variation of a traditional card game called pyramid solitaire. To play, you need a deck of regular playing cards. Remove the Jokers and picture cards. Then arrange twenty-one of the remaining cards in a six-row pyramid, like this:

Each row of the PYRAMID overlaps the preceding row.

The goal is to remove as many cards from the PYRAMID as possible. Two prerequisites must be met before removing a card. First, a card must be fully exposed. When the game commences, only the six cards in the bottom row meet this requirement. Second, you can only remove Tens or two fully exposed cards whose sum equals ten. In the PYRAMID above, you can take out three cards. The Seven of Clubs and the Three of Hearts can be removed, since their sum equals ten. You can take the Ten of Spades, since it's a Ten. After removing these cards, one new card, the Five of Spades, is fully exposed. (The Ace of Clubs is still overlapped by the Ace of Hearts.) There are now four exposed cards, but none of them can be combined to make ten. Therefore, you can't remove any other cards.

The game isn't over, though. Your playing deck consists of nineteen cards that aren't part of the PYRAMID. You turn these cards over, one at a time. When you turn a card over, you can use it to match any exposed card in the PYRAMID. Right now, if you turn over a Five, a Nine, an Eight, or a Four, you can make a match. If you turn over any other card, you can't form a match. Cards that don't form matches are put, face up, in a discard pile. You can go back to the discard pile any time you want and use the top card in the pile, but none of the other cards. The game ends when you've turned over all nineteen cards that weren't part of the original PYRAMID. Your score is the number of cards remaining in the PYRAMID. The smaller, the better. Three cards left is my lowest score of record. A second-grader named Sophie, however, once ended a PYRAMID with a single card.

In order to guarantee that PYRAMID will be fun and not a chore, I give the child a mathematical crib sheet.

$$1 + 9 = 10$$
$$2 + 8 = 10$$
$$3 + 8 = 10$$
$$4 + 7 = 10$$
$$5 + 6 = 10$$
$$6 + 5 = 10$$
$$7 + 4 = 10$$
$$8 + 3 = 10$$
$$9 + 2 = 10$$
$$+ 1 = 10$$

*Adding and Subtracting*

Every time a new card is exposed, I ask the child what number he needs to make a match. Scanning the list of equations makes the question easy to answer. The more a child plays PYRAMID, the more familiar these equations become, and the less dependent a child is on the crib sheet.

For a change of pace, switch the goal number for PYRAMID. Instead of trying to make ten, try to make nine. To play this version, first remove picture cards, Jokers, and Tens from the deck. Then give your child a math fact sheet listing equations that add up to nine. Set up your PYRAMID and you're ready to go. Next you can switch the goal number to eight. To do this, remove picture cards, Jokers, Tens, and Nines from your deck. This time, because the deck is reduced quite a bit in size, you need a smaller PYRAMID for the game. When going for eight, play with a five-row rather than a six-row PYRAMID. And give your child a fact sheet listing equations that add up to eight.

There are two ways to hold PYRAMID competitions. First, you and your child can take turns with the card deck and compare your scores round by round. Second, you and your child can each keep track of your best PYRAMID score to date and strive to top your own record every time you play. The second way allows your child to play by himself when you're too busy to join in.

*T*his game is for children who delight in difficult mathematical puzzles. Some children don't like this kind of work. Others love it. That's what makes horse races.

You play this game with an abbreviated deck of cards, meaning you have to remove all picture cards, Jokers, and Tens. When the deck is ready, deal five cards to each player. Using the numbers on these five cards, players try to make equations equal to ten. The player with the most equations wins. To make equations, you add and subtract any of the five numbers on your cards, but you can't use a number more than once in a single equation. That is, if you only have one 5 in your hand, you can't use $5 + 5 = 10$. Also, you can only use a given combination of numbers one time. If you've already used $6 + 1 + 3$, you can't include $1 + 6 + 3$.

Let's say you get this hand: Four, Six, Seven, Nine, and Ace (Ace equals one).

If you think hard, you'll discover four equations equalling ten: $6 + 4$, $9 + 1$, $9 + 7 - 6$, $7 + 1 + 6 - 4$. The first two equations are fairly easy for children to discover. The last two are tough. To find such equations, children need help. Your job is to provide as much help as needed.

Take Paul, for instance. Paul was not the best math student in the second grade; he was average. Still, he loved solving difficult math problems. On one particular deal, Paul wound up with Nine, Ace, Six, Three, and Eight.

## GRADES

two and three

## MATERIALS

a deck of cards
with picture cards, Jokers, and
Tens removed
paper
pencil

*Adding and Subtracting*

While he hunted for tens, I acted as recording secretary, writing every equation he discovered on a scorecard. It didn't take long for Paul to have his first success.

"Here's ten," he announced within half a minute. "I've got Nine and Ace—that's ten."

"Good for you," I said as I wrote the equation down.

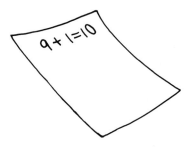

After this easy find, Paul was in for some hard work. He concentrated, but it was no use. As soon as Paul's gloomy voice declared, "I don't see any more," I offered some help.

"Try starting with the Six," I suggested. "Do you know what number you add to six to get ten?"

Paul used his fingers to figure this one out. "I need four to make ten, and I don't have a Four."

"Look carefully, Paul. Do you have any cards that equal four?"

"I see it," Paul said. "Three and Ace equal four. That means three and one and six equals ten." He doubled-checked this finding on his fingers. Sure enough, it worked. I wrote down Paul's second equation.

"You still have more equations," I said. "For instance, what does eight plus six equal?" As I talked, I pulled the Eight and the Six away from the rest of his cards.

"Fourteen," Paul replied after working out the problem. "But that's too big."

"I know, but what number do you take away from fourteen so it will equal ten?"

Paul needed four of my fingers to work out this problem. With our fingers all in a row, it was easy to see that lowering my four fingers would leave his ten. But Paul saw a new difficulty. He didn't have a Four. Then he remembered the Three and Ace. Paul started working this out. Six and eight is fourteen; fourteen take away one and take away three equals ten. Paul was grinning with pride as I wrote down this mammoth equation.

"I see two more equations," I said. "Do you want a hint?"

"Wait," he said. "Let me try." Paul started fiddling with the cards. He combined the Nine and the Three. "Is this right?" he asked.

"How much does nine plus three equal?" I asked in return.

"Twelve," answered Paul.

"How much do you take from twelve so that it equals ten?" was my next question.

Using ten of his fingers and two of mine, Paul concluded that twelve minus two equals ten. He didn't have a Two, so he started fiddling with the cards again. This time he tried the Three and Eight. Using his ten fingers and one of mine he discovered that eight and three equal eleven. Looking at our combined raised fingers, he observed that eleven minus one equals ten. What's more, he had an Ace in the hole. He was ready to dictate a new equation: eight plus three take away one equals ten. Paul had one more equation to find. He'd worked hard on the problem so far, and was understandably tired, so I gave him a clue.

"Nine plus eight equals seventeen," I said. "Can you figure out what you subtract from seventeen to get ten?"

Paul got a funny look on his face. He remembered that thirteen minus three is ten. He remembered that twelve minus two is ten. He remembered that eleven minus one is ten. Paul thought he detected a pattern.

"Is it seven?" he asked, a little haltingly.

"You got it," I replied.

"But I don't have a Seven," he complained.

"I know, but you have two cards that equal seven."

Paul looked over his cards. "Oh yeah, six and one are seven.

That means that nine and eight take away six and take away one is ten."

"That's amazing, Paul. That's the second equation you found that uses four numbers!"

This was the end of the line for Paul. Neither he nor I could find any more tens hiding in his cards. Now it was my turn to hunt for equations. Paul's score wouldn't be easy to beat, but I gave it my best shot. Paul took over the pencil and got a fresh sheet of paper for my scorecard while I took five cards from the deck. I wound up with Two, Three, Five, Five, and Nine. I studied my cards.

"Here's an equation," I said. "Five and five equals ten."

Paul objected that I was using the Five twice for one equation. I told him that in this case, since I actually had two Fives in my hand, it was a fair thing to do. Then I went on making equations. When I finished, I had five plus five, five plus two plus three, and nine plus three minus two. Paul carefully recorded each equation I discovered. When we compared scorecards, Paul had five equations while I only had three. Paul was delighted.

I was delighted, too. During the course of this game, Paul had done a lot of real mathematical thinking, the kind that makes for brain fatigue.

When your child is ready for more, try playing SIX CARDS MAKE TEN. Include just one extra card and get many more equations.

# chapter 6.

# Size and Shape II

There are five activities in this chapter. Two help children learn about geometry, and three help children learn to measure.

In 1975, Congress passed the Metric Conversion Act. This act says, in effect, "Come on, America, let's go metric!" Going metric will certainly be hard on those of us who, in our adult years, must adapt to a whole new measurement system. For children, however, it's a boon. For one thing, metrics is much easier than the traditional English system. What's more, metrics is the worldwide system of measurement—scientists and doctors converted to it long ago—and our children can only gain by joining in. But the general transition to metrics has yet to arrive. Most of our daily measurements are still of the inch, quart, and pound variety.

Here are three games that will make both the old and the new measurement systems feel like old hat to new mathematicians. IS IT? is a guessing game that demystifies centimeters, decimeters, and inches. SEWING PROJECTS sets children to cutting fabric the right size for a pillow or change pouch, thus giving them practical reasons to use rulers. COOKING gives children reasons to measure, too—this time in cups, liters, and milliliters.

Geometry activities begin with TANGRAMS. TANGRAMS are Chinese geometric puzzle pieces, three hundred years old, very useful, very geometric. ORIGAMI gives children the chance to learn about squares, triangles, rectangles, and quadrilaterals while folding paper into birds and cats.

All five of these activities require a certain amount of time on your part. You probably won't want to turn to these games after an exhausting day of work. Some Sunday, however, when you're relaxed and your child is bored, try opening *Games for Math* to this chapter. Scan the activities and pick out one that suits your mood and your child's creative nature.

# *IS IT?*

**GRADES**

one, two, and three

**MATERIALS**

paper
pencil
centimeter ruler

**W**hile my eyes were shut, Daisy took a centimeter ruler and carefully drew a three-centimeter line on a blank sheet of paper. She hid the ruler in her lap, then gave me permission to open my eyes. I looked at the line Daisy had drawn.

"How long is it?" Daisy asked.

I made my best guess: "Is it four centimeters?"

"No way!" Daisy said. She put the ruler alongside her line so I could see the correct measure for myself.

"I guess my eye isn't very accurate today. Now it's my turn to ask you an IS IT? question, and I'm going to make it tough. Without looking at the ruler, draw a line six centimeters long."

Daisy picked up her pencil and started drawing. When she finished, she stopped and considered her work. She decided her line was too long, so she erased a bit. When she finished erasing, we both wanted to know—is it a six-centimeter line? I handed her the ruler and she measured. Her line was just over six centimeters.

"That's excellent," I said with honest admiration. "You were so close."

"If only I'd erased a little more, I might have been just right."

"Being just right in this game is almost impossible. You're doing terrific when you're almost right. Now you can give me an IS IT? to try."

"OK. This time I want you to find something in the room that's fifteen centimeters long."

"Can I see the ruler before I hunt?" I asked.

"All right, you can look, but you can't touch."

I stared at the ruler for a few seconds. Then I began to walk around the room. Finally, I saw a book that seemed to be fifteen centimeters long. I handed it to Daisy and said, "I think this book is fifteen centimeters from top to bottom. Is it?"

Daisy measured the book. It was just over sixteen centimeters. We both agreed it was a good try.

What was Daisy learning while playing IS IT? She was developing firm notions about centimeters. Over the course of several months we played IS IT? four or five times, for about ten minutes at a stretch. In that short amount of time, Daisy internalized the length of a centimeter. She was remarkably accurate when estimating the length of a line, drawing lines of specific lengths, and estimating the metric length of everyday household objects. As a result of playing IS IT? centimeters held little mystery for Daisy.

She was also learning how to use a ruler. This is a challenging task, and I've found that children learn best when they begin with a modified ruler. The one I use is ten centimeters long and looks like this:

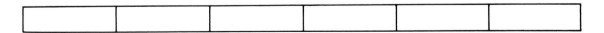

My ruler differs from the store-bought variety in two ways. First, I only mark centimeters. I leave off the smaller millimeter units. Second, there are no numbers. This means that children must count the distances they are measuring. They can't just read the numbers.

You can copy my ruler off this page and use it for playing IS IT? If you want to work with a longer ruler, then make two or three copies and tape them together. Ten centimeters equal one decimeter. Once your child is very familiar with centimeters, you might ask him to draw decimeter lines or search the house looking for things that are two or three decimeters in length.

In deference to us old people who grew up in a pre-metric world, you might want to play IS IT? in an inch version too. When Daisy and I play IS IT? with inches, I give her a modified inch ruler just like the modified centimeter ruler. The inches are marked but unnumbered, and the fractional markings are left off altogether. My inch rulers are six inches long.

# TWO SEWING PROJECTS

## GRADES

one, two, and three

## MATERIALS

sewing supplies
a metric ruler

**G**ive children practical reasons to measure and they'll work hard to master this mathematical skill. A bit of tailoring and upholstering, for instance. Here's a project for making a pillow.

## The Pillow

The idea is to produce a diamond-shaped pillow that measures twenty-six by twenty-two centimeters. You need a piece of material large enough to cut out two identical diamond shapes. Help your child measure and draw two diamond shapes on the back of the material. They should look like this:

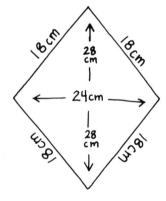

Next, help the child measure and draw a one-centimeter hem line all around the diamonds.

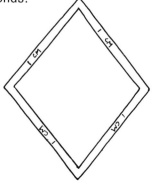

Cut out the shapes, and pin them together so the reverse sides of the material are facing out. Show your child how to stitch along the hem line. When he has sewn three sides and all but approximately six centimeters of the fourth side, he should turn the material right side out.

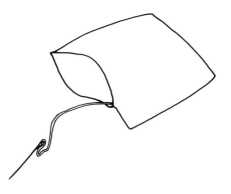

Now it's time to stuff the pillow. You can use commercial stuffing or cotton, or rip strips of old towels and other fabrics. The strips should be approximately two centimeters wide by ten centimeters long. When the pillow is full, stitch up the opening. That's all there is to it.

If your child would prefer a bigger pillow or a smaller one, a triangular or a square one, you can vary the instructions without much trouble.

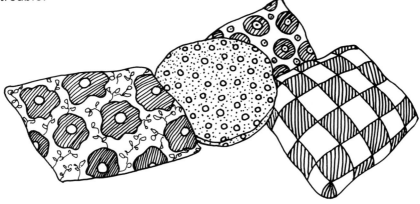

Here's a second project.

## Change Purse

I use felt to make the purse, though denim or any other heavy-weight cotton will do. Start by helping your child measure and cut a piece of material ten centimeters wide by twenty centimeters long.

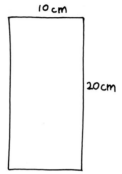

Now your child turns the material into a purse shape by folding up eight centimeters of material. This produces an eight-centimeter pouch topped by a four-centimeter flap.

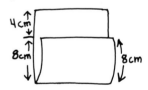

Help your child draw a one-centimeter hem on the left and right sides of the pouch.

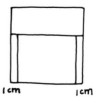

Show your child how to sew these hems. Then reverse the pouch so the stitching is on the inside.

On either side of the flap your child must fold inward and sew in place a one-centimeter strip of material.

Now the purse needs a Velcro fastener. If you have lengths of Velcro on hand, you can assist your child in measuring and cutting a six-centimeter piece of hook and a six-centimeter piece of loop. Then your child can sew the loop to the pouch and the hook to the flap. (Velcro sells waistband-size fasteners that will do for this job. If you use these strips, however, you lose a good opportunity to help your child measure.)

Is the Velcro in place? Then the purse is good enough to put your child's money in.

# MATH IN THE KITCHEN

**GRADES**

one, two, and three

**MATERIALS**

cooking supplies

**A**ll learning has its price. In the case of learning to measure capacity, the price will be paid by your kitchen floor. You will invite your child into the kitchen. You'll go to work on various recipes. And if the floor begins to resemble swampland, you can console yourself with the thought that although learning is usually invisible, now you can see it well.

Should you teach your child the metric system (liters and milliliters) or the English system (cups and pints)? I say: Both.

Here's a recipe using the English system:

**Peanut Butter Balls** (makes about fifty balls)

Ingredients:
2 cups peanut butter
1 cup rolled oats
1 cup powdered milk

1 cup raisins
¼ cup honey

1. Mix all the ingredients together in a bowl.
2. Form the mixture into teaspoon-size balls.

You can always find other English-system recipes in a cookbook. Just look for simple ones.

The metric system requires metric measuring cups, which have become easy to find at supermarkets and houseware stores. It might also be a good idea to get a set of gradated milliliter spoons. But if you can't put your hands on metric equipment, you can still cook metric by relabelling your old cooking equipment according to the following rules:

250 milliliters equals one cup
125 milliliters equals ½ cup
80 milliliters equals ⅓ cup
60 milliliters equals ¼ cup
15 milliliters equals one tablespoon
5 milliliters equals one teaspoon
2.5 milliliters equals ½ teaspoon
1.25 milliliters equals ¼ teaspoon

Suitably equipped, you're ready to prepare recipes like these:

**Elizabeth's Gooey Chewies** (makes one panful)
Ingredients:

1 egg
160 ml brown sugar, packed
80 ml maple syrup
125 ml flour

1.25 ml salt
1.25 ml baking soda
180 ml chopped walnuts
60 ml chopped dates

1. Preheat oven to 350°
2. In a mixing bowl, beat the egg.
3. Add brown sugar and maple syrup.
4. In a second bowl sift the flour, salt, and baking soda.
5. Blend the wet and dry ingredients.
6. Add walnuts and dates.
7. Pour into a well-greased 8″ x 8″ pan.
8. Bake at 350° for 25 minutes.
9. Let cool so that Gooey Chewies can harden a bit.
10. Cut up into small squares—they're very sweet!

**Banana Bread Surprise** (makes one loaf)

Ingredients:

3 bananas
60 ml honey
180 ml sugar
45 ml butter, softened to
   room temperature
2 eggs
2.5 ml vanilla extract

500 ml flour
2.5 ml salt
2.5 ml baking soda
5 ml baking powder
100 ml strawberry jam, raspberry
   jam, or orange marmalade

1. Preheat oven to 350°.
2. Sift together flour, salt, baking powder, and baking soda.
3. In a different bowl, mash bananas; then mix in sugar, butter, eggs, honey, and vanilla.
4. Blend dry and wet ingredients together.

5. Pour about two-thirds of the batter into a well-greased bread pan.
6. Gently spread a layer of jam over the batter.
7. Pour on remaining batter.
8. Bake at 350° for 50 to 55 minutes.

**Vegetable Soup** (makes six to eight servings)

Ingredients:

30 ml olive oil

750 ml chopped vegetables:
   zucchini, tomatoes,
   carrots, and celery

1500 ml water

6 bouillon cubes

1 bay leaf

a dash of oregano

salt and pepper to taste

250 ml elbow macaroni

150 ml drained chick peas

1. In a large pot, sauté vegetables in oil for about 15 minutes.
2. Add water, bouillon cubes, oregano, bay leaf, and macaroni.
3. Increase heat and boil for 5 minutes.
4. Reduce heat to a simmer and add drained chick peas.
5. Simmer for 30 minutes.
6. Add salt and pepper to taste.

If you get into the habit of using metric vocabulary as you cook—"We need 125 milliliters of sugar. Do you want to pour?"—you'll help your child feel at home with both metrics and measuring. That's a good idea for practical skill, and a good idea for conceptual development.

**W**hat's a TANGRAM puzzle? Here are two classic—and classy—examples:

## TANGRAM Cat Puzzle

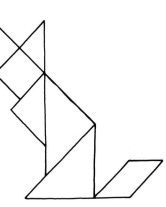

**GRADES**

one, two, and three

**MATERIALS**

cardboard, poster board, or oaktag

paper

pencil

tape

scissors

large envelope

## TANGRAM Boat Puzzle

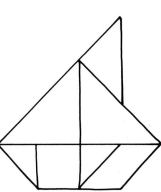

TANGRAMS were invented in China, probably in the seventeenth century. Lewis Carroll studied them. Edgar Allan Poe owned a TANGRAM set made of ivory, with inlay carvings of exotic birds, flowers, and Chinese villagers in a garden. And if TANGRAMS proved pleasing to the authors of *Alice in Wonderland* and "The Murders in the Rue Morgue," you may confidently assume that something about this three-hundred-year-old puzzle is good for the imagination.

It's also good for geometric thinking.

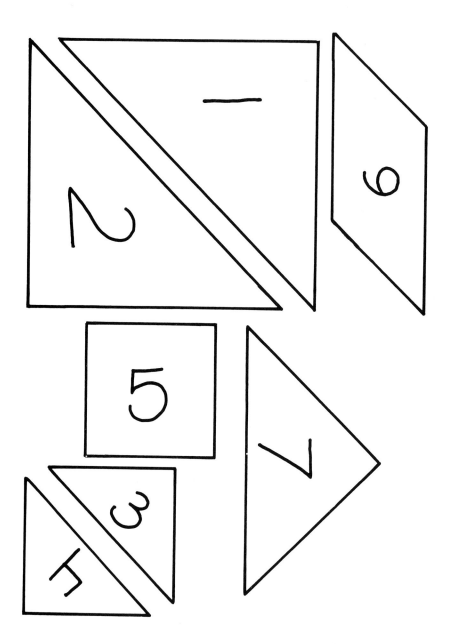

A TANGRAM set is made up of seven pieces: one square, one parallelogram, one medium-size triangle, two big triangles, and two little triangles. You'll need to make your own set of pieces, which you can do by photocopying the drawings on the facing page, or by carefully tracing them, with the help of a ruler. Tape the designs to cardboard or oaktag and cut out the shapes. Number each piece 1 through 7, just like the models. You can use the pieces with the numbered side up       or down:

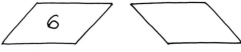

You'll want to store the pieces in an envelope.

When you introduce your child to TANGRAMS, it's best to begin with simple puzzles. Start by taking pieces 3 and 4 out of the envelope. Have your child close his eyes, place the pieces on a sheet of paper, and arrange them in a square shape.

Next, trace the square. You may want to tape the pieces together so they don't slide when you trace. Now pick up the pieces and carefully remove the tape. Tell your child to open his eyes, hand him the puzzle pieces, and dare him to recreate the square with TANGRAMS.

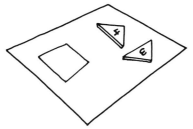

If your child succeeds in making the square, he'll probably want to tackle another puzzle. Tell him to close his eyes again. Then take the same two pieces, make a new design, trace it, and see if your child can solve the puzzle. Here are some designs I came up with using just pieces 3 and 4:

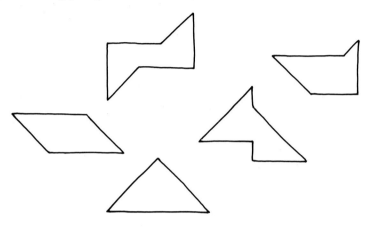

And their solutions:

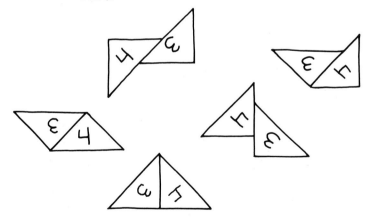

Pieces 3 and 6 make for other possibilities:

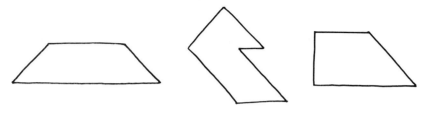

And the solutions:

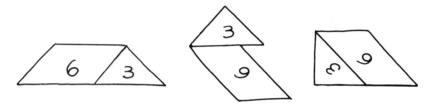

With pieces 3, 4, and 6 you can make these puzzles:

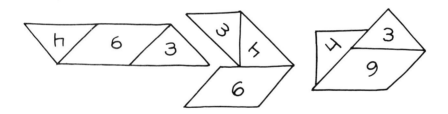

With pieces 3, 4, and 7 you can make:

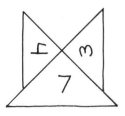  

When you include all seven pieces you can make:

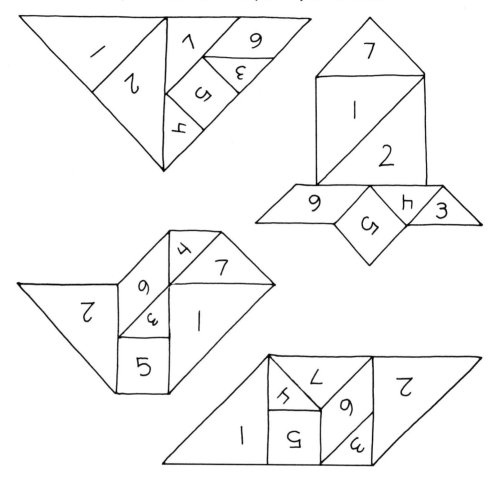

If your child has trouble solving some of these puzzles, give hints. Telling your child that the triangle is in the right place but the square isn't will help him come up with a solution before he gets too frustrated.

Try for yourself. Close the book, and try making a square using all seven pieces. Oooh, that's tough. Would you like a hint? The solution to this puzzle is at the end of this activity. Go ahead and take one peek at it. Then close the book and try again.

When your child solves TANGRAM puzzles, he goes through the same process you just did. First, he exercises his spatial imagination. Children with a firm grasp of space and spatial relationships have a real advantage when it comes to studying geometry. Second, he learns quite a lot about shapes. Just by fiddling with pieces 1 and 2, your child may discover that two triangles can make a square, a parallelogram, or another triangle. The more your child fiddles, and the more pieces he fiddles with, the more discoveries he will make.

TANGRAMS are more fun if you make two sets, so that you and your child can both work at the same time on designing puzzles and solving them. Of course this means you must try and solve the puzzles your child creates. Expect no mercy from your child. He will surely create the trickiest, toughest, most puzzling puzzles he can. And he isn't likely to give you any helpful hints, either.

**Solution to the seven-piece square puzzle:**

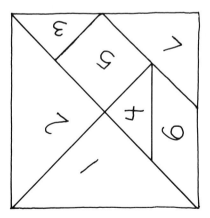

# ORIGAMI

**GRADES**

one, two, and three

**MATERIALS**

paper
scissors
crayon

*T*he philosopher and mathematician René Descartes conceived of new geometric ideas while watching a fly zigzag across the ceiling. Mathematical learning takes place in unusual ways. Your child can learn a lot by building a paper cat, dog, or bird. Just follow the rules or ORIGAMI, the Japanese art of paper-folding. The result? Discoveries in geometry.

To make ORIGAMI animals, your child will start out with a square piece of paper, fold it in half from point to point, and end up with—a triangle.

Next your child folds back the tip of the triangle and ends up with—a trapezoid.

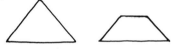

Children who design lots of paper animals begin to notice that half a square equals a triangle and that a triangle without a tip equals a trapezoid.

When making ORIGAMI figures with children, I don't ram geometry facts into their heads. I assume that with enough playful encounters children will sense these relationships for themselves. And I've noticed that children who have lots of experiences with ORIGAMI (and TANGRAMS, too) do develop better instincts for geometry.

All ORIGAMI figures begin with square sheets. You can buy specially prepared ORIGAMI paper in art supply stores and some toy stores. But you don't need special paper. Typewriter paper cut into squares works very well.

The easiest way to square a rectangular sheet of paper is to fold one corner of the rectangular sheet so that you form two overlapped triangles with a strip of paper at the bottom.

**106**

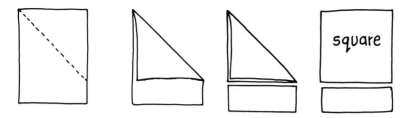

Cut off this strip, and you're left with a square sheet. You can use this square as reference for cutting squares from other sheets of typing paper.

And now, take some paper, and fold as you read.

## Make a Cat

1. Start with a square sheet of paper. Fold the square to make this triangle:

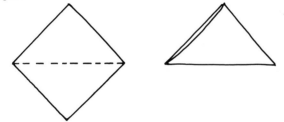

2. Fold the top point of the triangle down and forward.

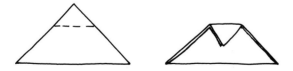

3. Fold the two bottom points up and forward.

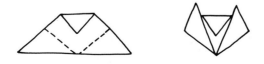

4. Turn the paper around and draw cat features on the pentagon-shaped face.

## A Dog

1. Start with a square sheet of paper. Fold the square to make this triangle:

2. Fold the top corners of the triangle down and forward. Now you have a diamond shape bordered by two triangular flaps.

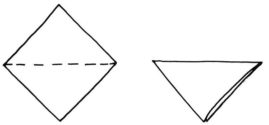

3. Fold the top and bottom of the diamond down and backward.

4. Draw a dog face.

## A Bird

1. Hold a square sheet of paper like this:

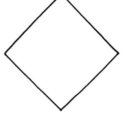

2. Make triangular forward folds on either side, thus changing the square to a kite.

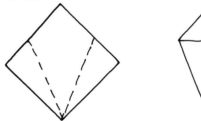

3. Make two more triangular folds, changing the kite into a diamond.

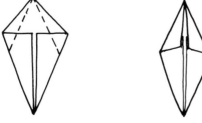

4. Turn the diamond (clockwise) on its side and fold the bottom half back. Now you have a triangular shape.

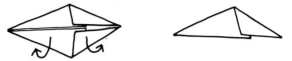

5. You now have the bird's body, and it's time to create a neck. Make a triangular bend in the front of the bird and fold the triangle up: to make this fold, the bottom of the neck should turn inside out as it tucks up between the two sides of the bird's body.

6. Now you need a head. Make a triangular bend at the top of the neck and fold it down, tucking the head in between the two sides of the bird's neck.

A cat, a dog, a bird—they're more diverting than a ruler and graph paper. And they're no less useful for teaching fundamental geometric shapes.

# chapter 7.

# *Multiplication and Division*

I announced the subject of today's lesson and Samantha said, with pride and satisfaction at the step she was about to take, "Finally —multiplication."

To Samantha, multiplication represented a major educational landmark, which she had been eagerly anticipating, and a wee bit fearing.

In fact, though, Samantha was already well on her way toward expertise without knowing it. For several weeks, I'd been teaching her the concepts behind multiplication—and division, too. Only instead of using the words and symbols we attach to those operations, I'd been teaching her STAR COUNT, VICTOR VAMPIRE'S BIRTHDAY, LOTS OF BOXES, and CALCULATING MATH. These games do without $\times$ or $\div$, but when you play, you're operating by the rules of $\times$ and $\div$ nevertheless.

Schools usually introduce multiplication and division sometime toward the end of second grade. It isn't until third grade, however, that the real push begins. In third grade children not only learn the meaning of these operations and how to solve the problems, they also start memorizing those tables. Usually this work continues well into fourth grade, and many children struggle with tables right through fifth and sixth grade.

The basic concepts that underlie multiplication and division are, however, well within the grasp of first-graders—if they draw pictures or count real objects rather than try to solve equations. That's the beauty of STAR COUNT, VICTOR VAMPIRE'S BIRTHDAY, and LOTS OF BOXES. These games are so easy to play that first-graders can go a few rounds, and yet these games are perfectly suited to second- and third-graders engaged in more formal study of multiplication and division.

CALCULATING MATH is more advanced, and you shouldn't bother with it until your child is in the middle of second grade.

The idea in these four games isn't to get children to memorize their multiplication and division tables (although one variation of CALCULATING MATH can do just that). The purpose is to familiarize children with the fundamental concepts underlying multiplication and division. It takes practice to master these ideas. Here are four ways to practice playfully. (And don't forget the skip counting games in Chapter Four. They help children with multiplication, too.)

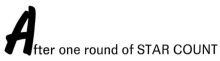

**A**fter one round of STAR COUNT my scorecard looked like this:

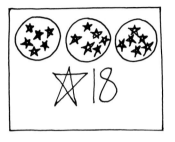

**GRADES**

one, two, and three

**MATERIALS**

paper
pencil
one playing die

Elise, a first-grade math fan, had a scorecard that looked like this:

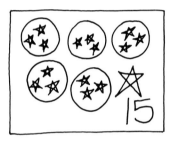

Elise let out a moan. "You won. You have eighteen stars, and I only have fifteen."

"It's only fair," I said. "You won the last round. That means you've won once and I've won once. Let's see who can be first to win five rounds."

Elise was game. She rolled the playing die and got a three. This obliged her to draw three circles on her paper scorecard.

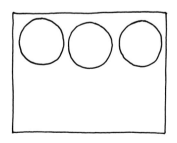

She threw the die again and got a five. That told her to draw five stars inside each circle.

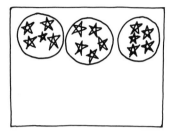

She counted all her stars.

"I have five and five; that's ten. Then I have eleven, twelve, thirteen, fourteen, fifteen. Hey, that's the same as I had before." She drew a larger star and wrote her score alongside.

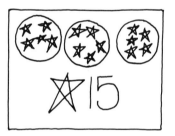

Next it was my turn. I rolled a six and drew six circles.

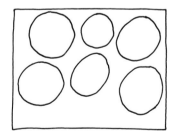

"You're going to win again," grumbled Elise. "You have so many circles."

"Maybe, if I'm lucky, I'll win. I hope I'm lucky. What do you hope?"

"I hope you're unlucky."

On that encouraging note, I threw the die. I got a one. I drew a single star in each of my circles and added up my score.

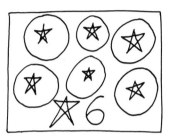

"Only one star in each circle. I have six circles, and I have six stars. Phooey."

"I won! I won!" Elise shouted.

"You won this time," I chided. "I may be luckier the next time. Do you want to go first or should I?"

"Me!" Elise said.

What do students learn when they play STAR COUNT? They get a kind of preliminary, pictorial version of how teachers describe multiplication. Teachers say that multiplication is a way of grouping numbers. You can think of the multiplication equation $3 \times 4 = 12$ as three groups with four in each group. STAR COUNT makes a perfect picture of that explanation.

Alternatively, teachers describe multiplication as repeated addition. In this explanation, $3 \times 4 = 12$ is expressed as $4 + 4 + 4 = 12$. STAR COUNT gives children a graphic demonstration of that multiplication explanation as well.

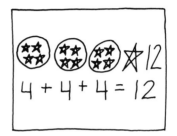

Children who play STAR COUNT may make interesting discoveries about multiplication. In this particular game, Elise noticed that three groups of five stars is the same number as five groups of three stars.

STAR COUNT is simple enough for first-graders. But the concepts it exemplifies are sophisticated enough to be useful to third- and fourth-graders. By the way, if your child has trouble drawing stars, you can substitute *x*'s, or asterisks, or anything.

**A**lthough Gregory was a first-grade math whiz, he wasn't ready for formal lessons in multiplication or division. He was, however, ready to play with higher mathematics informally, through stories. The first story I told Gregory starred a vampire named Victor.

"Victor Vampire was planning his birthday party. He needed enough food for himself and his three friends, Roger Werewolf, Gloria Ghost, and Zee Dee Zombie."

As I talked I made little drawings of Victor and his guests.

"Victor wanted everyone at the party to have three eyeballs for appetizers. Can you draw three eyeballs next to each guest and tell Victor how many eyeballs to buy at the Ghoul Grocery Store?" I asked.

"Sure," replied Gregory. First he drew.

Then he counted. When he got to twelve, he stopped. "Twelve," he said. "He needs twelve eyeballs—ugh!"

I went on, "Victor also wants blood cocktails for everyone. He knows he needs at least five glasses of blood for each person. Can you figure out how much blood he needs? You can draw five little glasses in front of each character if you like."

Gregory liked.

"I can count these easily," he bragged. "I can count by fives: five, ten, fifteen, twenty. He needs twenty glasses."

"How clever to think of counting like that. Now here's a really tricky problem for Victor. His mother, Violet Vampiress, sent him a special birthday treat—sixteen candy-coated monkey guts. Victor wants to share these treats equally with his friends. How many should he give to each person so that the guts are equally divided?"

"I don't know how to figure that out," complained Gregory.

"Why don't you start by drawing the sixteen monkey guts?" I suggested.

"What do monkey guts look like?"

"I'm not sure. But if you draw sixteen little squiggles, we'll call them candy-coated monkey guts."

"Now, one at a time, draw a line from each gut to each guest at the party. Give Victor a gut, then Roger, then Gloria, then Zee Dee, then give Victor another gut, then Roger, and so on. Keep going until you've given away all the guts. Then count how many treats each guest receives."

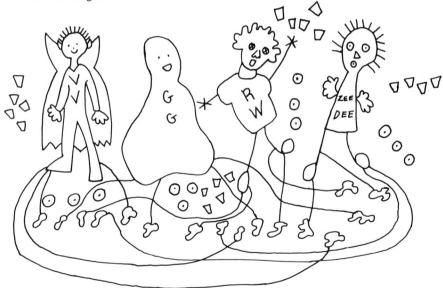

"They get four guts apiece," Gregory declared after completing this work.

"Right you are."

To solve these problems, Gregory added and counted. He didn't multiply or divide. Still the story of Victor Vampire's birthday party prepared Gregory to think about multiplication and division. A few years from now when Gregory's teacher shows him how to record three eyeballs to four party guests as $3 \times 4 = 12$, or sixteen monkey guts shared four ways as $16 \div 4 = 4$, he may find this new mathematics easy to understand.

Ironically, these stories presented bigger problems for me than for Gregory. I had to come up with tales to tell. Fortunately, I've found that even flimsy stories capture young imaginations. To help you tell stories to your child, here are a few starter tales you might try.

Don't forget to accompany your storytelling with illustrations, which can only make the stories more amusing—and easier for the budding mathematician.

★

Three clowns performed with the Hunky Dunky Circus. One day these clowns decided to play tricks on the other circus performers. They sneaked up on the elephant trainer and took away all his peanuts. Each clown took five bags of peanuts. How many did they take from the elephant trainer? Then they went to the trapeze flyers' tent. The clowns decided to put roses in the trapeze flyers' shoes. They found six shoes in the tent and put two roses in each one of the shoes. How many roses did they use? The clowns had nine candy canes. They decided to eat these candy canes up, but they wanted to share them evenly. How many candy canes did each clown get to eat?

★

Six Martians came to visit Earth. They wanted to find out about Earthly things. The first place they saw was a candy store. "Let's try some bubble gum," Mickey Martian suggested. "Give us two packages apiece, please." How many packages of gum did they buy?

"That will be three dollars," said the store owner.

Morris Martian handed the owner ten dollars. "Keep the change," Morris said. How much extra money did the store owner get from Morris?

Next Mindy Martian saw a pet store. In the window were eighteen kittens. "Let's take these kittens back to Mars," she said. "We can share them." If they share the kittens evenly, how many will each Martian get?

The Martians saw an ice cream store. "Oh, that looks delicious. I could eat five cones all by myself," said Margaret Martian. All the other Martians felt exactly the same. How many cones did the Martians buy?

After this they felt a little funny in their tummies. Each Martian had three tummies. How many stomachaches did these Martians have?

"I've had enough of Earth," said Martha. "Let's go home." So away they all flew.

★

Mother Rabbit went shopping for her five babies. She had very sweet babies, but one thing made them mad. If one baby bunny got more than his equal share, all the other babies stamped and kicked. The first store Mother Rabbit visited was a toy store. She saw a teddy bear. "That bear would be a nice present for my biggest bunny, Teddy." Then she thought, "I'd better get five of these bears or my other bunnies will be angry." Each teddy bear cost four dollars. How much money did Mother Rabbit spend in the store?

Next Mother Rabbit went to the grocery store. She

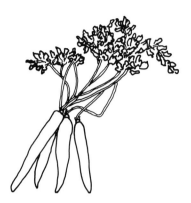

saw a beautiful bunch of carrots. She counted fifteen carrots in the bunch. Mother Rabbit thought to herself, "Thank goodness there are enough here so my bunnies can share the carrots equally." Is Mother Rabbit right, or will there be a fight when she gets home?

Finally, Mother Rabbit went to buy her baby bunnies some winter ear warmers. She needed two ear warmers for each bunny. Mother Rabbit went into a hat store and said, "May I have twelve ear warmers please?" Is this the right number or the wrong number of ear warmers for her babies? How many warmers should Mother Rabbit buy?

★

What should you do when your child has trouble solving a number-story problem? Help him out. You can make suggestions about drawings. You can restate the problem, making the solution more obvious. Then, make sure the next problem is easier to solve.

What if your child likes the stories but hates to draw? Do the drawing yourself. Don't worry about the artistic merit of these sketches. Any old stick figure will do just fine.

What if your child dislikes the whole thing—pictures, stories, problems, all of it? Then forget number stories, at least for the next six months.

# M

ost children get a kick out of drawing on graph paper. That's the secret behind LOTS OF BOXES. The best kind of graph paper for this game is the ¼ inch square variety available in any well-stocked stationery store. But any kind of graph paper will do.

The idea is to make a bigger rectangle than your opponent. You make rectangles by throwing the die. Your first throw determines the length of your rectangle. If you throw a four, your rectangle is four squares long.

**G R A D E S**
one, two, and three

**M A T E R I A L S**
graph paper
pencil
one playing die

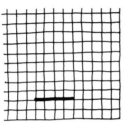

Your second throw determines the height of your rectangle. If you throw a three, your rectangle is three squares high.

Now you have enough information to finish off the rectangle.

This rectangle consists of twelve little boxes. That means your score for this round of LOTS OF BOXES is twelve. Write your score underneath the rectangle.

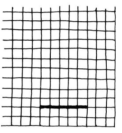

The child goes next. He throws the die once and finds out how long his rectangle will be.

He throws the die a second time and finds out how high his rectangle will be.

Then he completes the rectangle, counts the little boxes, and records his score.

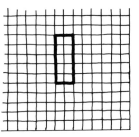

The child wins, twenty to twelve. That's all there is to LOTS OF BOXES.

One of these days, when you're playing, you can show your child the mathematical way to write his LOTS OF BOXES score. When your child creates this box,

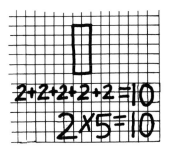

record his score by writing $2 + 2 + 2 + 2 + 2 = 10$. Tell the child to read this math sentence as follows: two boxes add two boxes add two boxes add two boxes add two boxes equals ten boxes altogether. Then show him an alternative way of writing his score: $2 \times 5 = 10$.

Tell him to read this math sentence as follows: two boxes five different times equals ten altogether. Then announce with a bang that $2 \times 5 = 10$ is a multiplication equation! Wow, multiplication —that's amazing!

Of course you can also record the score for this picture as $5 + 5 = 10$. That's five boxes going down plus another five boxes going down equals ten boxes. The multiplication equation for this is $5 \times 2 = 10$, or five boxes in two different columns equals ten boxes altogether. The choice is yours.

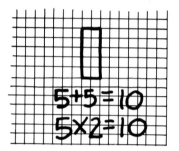

Some children get such a kick out of this discovery that they want to write multiplication equations for each box in the game. Other children couldn't care less about expressing game results in multiplication equations. If your child doesn't want to write his score in multiplication-eze, that's OK. But you can write your score in this fancy mathematical way, even if your child chooses not to.

**M**atthew was finishing a very successful second-grade year. Throughout the school year, math was his favorite subject. He could add in a jiffy. His subtraction was spiffy. Now he was eager to multiply and divide. I presented these new mathematical operations to Matthew by playing CALCULATING MATH.

"I have a calculator game today," I said. "To start this game, I make the calculator count by threes. To do that, I press the 'plus' button and then the 'three' button. See how three pops up on the calculator display."

**GRADES**

two and three

**MATERIALS**

a calculator
(one with an automatic constant for addition and subtraction)

"Next I press 'equals.' The first time I press 'equals,' the three stays on the display. But when I press 'equals' again, the calculator adds three, and a new number shows up. Can you predict what that number will be?"

"That's easy. Three and three are six."

"Are you sure?"

"Sure," he said with some annoyance in his voice.

"OK," I said. "Press the 'equals' button and prove it." Matthew pressed the button, and got a confirmation. (If your calculator doesn't do this, you can't play this game.)

"You did it. You out-calculated me! Now, can you predict what number will pop up on the display when you press 'equals' again?"

Matthew looked at the calculator and said, "That means six plus three, right? So it must be nine."

"Go ahead and press," was my sole reply.

We continued adding threes, and the calculator showed nine, twelve, fifteen, eighteen, twenty-one, and so on, until thirty. At first, Matthew could add three to the number on the display and figure out what would pop up next. When the numbers got larger, he counted using his fingers as an aid. Usually his predictions were accurate, but a couple of times they weren't. Matthew got nervous when he made mistakes. After one foul-up he even wanted to quit the game. A little reassurance from me, however, and he was ready to go on. When Matthew completed this count-by-three climb, he agreed to take on a new challenge.

"Let's go again, but let's do things a little differently. You'll count by threes again, so I'll press the 'plus' button and the 'three' button. Only this time I want you to predict what number will show up on the display after you press 'equals' four times. Remember, the first time you press 'equals' the display will say three. You need to figure out what it will say after the second time you press, then the third, then the fourth. Do you think you can do that?"

"I think so," he said gravely. "Just give me a chance, I'll figure it out." He fiddled with his fingers, he mumbled various numbers and finally declared, "Twelve. It'll say twelve."

"Well, there's only one way to find out," I said. "Go ahead and press the 'equals' button four times."

Matthew pressed and as the display changed he chanted along, "Three, six, nine, twelve! I told you so!"

This calculating was hard work, and Matthew's face radiated the pleasure of mastering a difficult job.

For the next ten minutes, I asked Matthew more calculator questions.

"Counting by sixes, what will the display show once you've pressed 'equals' three times?"

"What will the display show if you press 'equals' once more?"

"Can you count by fours up to twenty? Go one number at a time and we can check your count on the calculator display."

The next time we played with the calculator, I gave Matthew some new and harder jobs to tackle.

"This time I am going to begin by pressing the 'nine' button. That makes nine show up on the display.

"What do you think will happen if I press the 'minus' button, then the 'three' button, and then the 'equals' button?"

"Does that mean nine take away three?" Matthew asked.

"It sure does," I answered.

"Nine take away three is six," proclaimed Matthew.

"Good," I said while pressing the buttons and making six appear. "When I press 'equals' again, the calculator will subtract another three. What number do you think will show up then?"

"Six take away three makes three," Matthew answered.

"And if I press 'equals' again?"

"It'll say zero."

"Super calculating, Matthew. This time see if you can calculate in your head. First, I'm going to start the calculator on fifteen. Then I'm going to tell it to subtract three. How many times will I hit 'equals' before I get to zero?"

"I don't know if I can do that," said Matthew. "It's too hard."

"I know it's hard, but I think you can do it. Start by figuring out fifteen minus three."

Matthew started counting backwards from fifteen. He got to twelve. I told him to subtract a second three. This got him to nine.

I told him to subtract a third three. That's when Matthew interrupted to announce that he could do the rest by himself, thank you. After considering the problem a bit longer, Matthew smiled.

"You press 'equals' five times. Am I right? Is that it?"

"Let's check it out and see what happens," I replied. After pressing "equals" five times, Matthew saw zero and grinned from ear to ear.

"Do you think you can do another backwards count?" I asked.

"I'll try," he answered.

"OK, if I start the calculator on sixteen and I press 'minus' and 'four,' how many times will you press 'equals' before you reach zero?"

With a little bit of help, Matthew answered this question correctly. Then I asked one more question: "If the calculator starts at twenty-four and I press 'minus' and 'six,' how many times will you press 'equals' before you reach zero?"

The whole while we played CALCULATING MATH, I never said the words multiplication or division. The symbols $\times$ and $\div$ were never used. I hoped that by playing CALCULATING MATH Matthew would all by himself discover the link between addition and multiplication, subtraction and division. Happily, that's what happened. After a few weeks of playing CALCULATING MATH, I gave Matthew a formal classroom-type explanation of multiplication and division. I showed him that three times eight equals twenty-four, or $3 \times 8 = 24$, is the same as saying that adding eight three times equals twenty-four, or $8 + 8 + 8 = 24$. Then I showed him that fifteen divided by three equals five, $15 \div 3 = 5$, is another way of saying that you must subtract three from fifteen five different times to get to zero: $15 - 3 - 3 - 3 - 3 - 3 = 0$. As I was explaining, Matthew blurted out, "Hey, that's what we did with the calculator. I know how to do that! You mean that's division? I get it now."

A year later, in the middle of third grade, Matthew faced the task of memorizing his multiplication tables. Again, I pulled out the calculator.

"You're doing a great job learning your tables, Matthew. Do you think you can multiply as well as a calculator? Pick out a table and we'll see how you do."

"I want to try the fives table. That's a really easy one."

"OK, the fives table it is. I'm going to press plus five. Now you start your table. Every time you say a number I'll press 'equals.' If you are right, your number will show up on the calculator display. If you are wrong, we can start the fives table over a second time. Are you ready to go?"

Matthew started reciting: five, ten, fifteen, twenty . . . After he got to fifty, we stopped.

"You did that beautifully. In fact, you say your fives table faster than I can pop them on the calculator. Do you think you can do as well with a harder table? How about the fours? If you run into trouble, if you can't remember what number comes next, you can always add four to the number on the display. Add four, add four, add four—that's multiplication."

Matthew nodded his head, took a deep breath, and began: "Four, eight, twelve, sixteen . . ."

# chapter 8.

# The Number System

Two ideas lie at the heart of our number system. The first idea: our number system is based on ten. We use ten symbols to write numbers: 0, 1, 2, 3, 4, 5, 6, 7, 8, 9. We group tens together. Twenty means two groups of ten. Thirty means three tens. Ten tens makes a hundred, ten hundreds makes a thousand, and onward thus to millions, billions, and trillions. Because ten has this special mathematical role, mathematicians call our number system a base ten system.

The second idea: a numerical symbol changes value depending on where you write it. Consider these numbers: 2, 21, 236, and 2,678,350. The symbol 2 appears in each number, but with a different meaning each time. In 2 it means two ones. In 21 it means two tens. In 236 it means two hundreds. In 2,678,350 it means two millions. We call this special characteristic of numbers place value, since placement determines value.

Base ten and place value are difficult ideas to grasp, even for adults. Yet we introduce this bit of math theory to children sometime toward the end of first grade and hope the ideas are firmly entrenched by the end of third grade. Why do we introduce these ideas so early? Because addition and subtraction with numbers larger than ten are a lot easier if you have some understanding of the number system. Most teachers know, however, that even gifted students need two or three years to absorb these ideas.

Is it worth spending so much time on place value and base ten? Absolutely. When fourth-, fifth-, and sixth-graders run into trouble with math, you can bet their problems stem from confusion about how the number system works. If you help these children comprehend the number system, you discover that many of their problems in addition, subtraction, multiplication, division, decimals, and fractions start to clear up simultaneously.

In this chapter, you'll find six games that make these profound mathematical ideas a little easier to fathom. Two games, A BUNCH OF BEANS and FIFTY WINS, help children understand the special role ten has in our number system. Three games, BAG OF CHIPS, THROW A NUMBER, and THREE POTS, help children understand place value. The last game, GROUP TEN, helps children understand how base ten and place value work together when you add and subtract large numbers.

It's best to wait until your child is in second grade before playing any of these games at home. If you wait, your child's teacher will have laid the mathematical groundwork necessary to understand the games. Then all you need do is play.

# A BUNCH
OF BEANS

**GRADES**

two and three

**MATERIALS**

a bowl filled with at least
two hundred dried kidney beans

transparent tape

paper and pencil to make
your own game board

a deck of playing cards with
picture cards and Joker removed

**A** BUNCH OF BEANS helps second- and third-graders understand the complicated role of ten in the number system. Just think about twenty-seven. What does twenty-seven mean? It means twenty-seven individual numbers, and it *also* means two groups of ten and seven more. Thirty-six means three groups of ten and six more. Seventy-eight means seven groups of ten and eight more. It's hard for children to think of seventy-eight in this way, but playing A BUNCH OF BEANS makes it a little less hard.

You need a bowl filled with at least two hundred dried kidney beans, a deck of playing cards without picture cards or Jokers, and a playing board for each player. The playing board has one hundred little ovals organized in ten rows of ten ovals each. Draw it to look like this:

Rules for A BUNCH OF BEANS are very simple. You and your child take turns drawing cards from the abbreviated card deck. The number on the card tells you how many beans to take from the bowl. First hold the beans in your hand, then place them on the board, one bean to an oval. The first player to fill up his board with one hundred beans wins the game.

As your child plays A BUNCH OF BEANS, he may notice certain interesting things about numbers. Look at this game board:

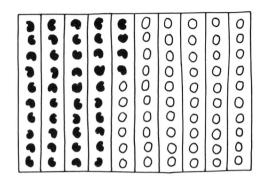

What's the score? To find out, you can count individual beans, or you can count groups of ten and then count the remaining beans. It doesn't take many rounds of A BUNCH OF BEANS before children discover the advantage of counting by ten: ten, twenty, thirty, forty, and then counting the remaining beans one by one: forty-one, forty-two, forty-three, forty-four.

Look at these two game boards:

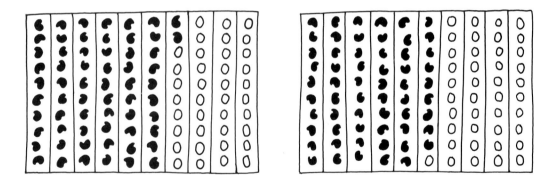

Who is winning? There are two ways to find out. You can count individual beans, or you can compare groups of ten. The quickest way to tell is by comparing tens. The player on the left has six tens. The player on the right has five tens. The player with six tens is winning. Not all children see things so clearly, however. Some children focus all their attention on the last column of beans. Since the player on the right has nine beans in this column and the player on the left has only two beans, these children believe the player on the right is winning. If your child has this problem, gently focus his attention on the *filled* columns of beans. Then ask, "If you count all the beans on my board and all the beans on your board, who do you think will have more?"

If this question alone doesn't change your child's perception, you will just have to count all the beans on each board.

Finally, look at this board:

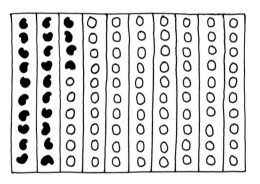

Now picture the board with ten more beans:

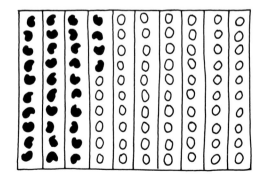

Although the score has changed, something is the same. Look at the right-hand column of beans. That column looks exactly the same as before. Remarkably, whenever a player adds ten beans to his board, the number of beans in his right-hand column remains unchanged. This is a reliable pattern. Children can take advantage of this pattern whenever they want to add ten to a number: twenty-four add ten is thirty-four, sixty-one add ten is seventy-one, forty-six add ten is fifty-six. If children play A BUNCH OF BEANS a bunch of times, they may discover this pattern. Then they'll never again need fingers to add ten.

If your child likes A BUNCH OF BEANS you might try playing the game in reverse. Start out with one hundred beans on your board. Then pick cards to find out how many beans you remove. The first person with an empty board wins the game.

A BUNCH OF BEANS is designed to help children make mathematical discoveries. It's also designed to be a good game. If you lecture your child about counting by tens or adding ten to a number, the game will be spoiled. Why not leave lectures to the teachers? Opt for playful ways to help your child instead.

# FIFTY WINS

**GRADES**

two and three

**MATERIALS**

paper
pencil
a bowl of at least one hundred
dried kidney beans
oaktag or poster board
paper fastener
scissors

**F**IFTY WINS is another game that helps children appreciate the importance of ten in our number system. You need to draw two game boards that look like this:

 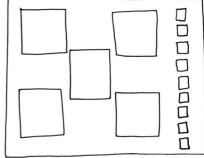

The board should be big enough so that each little square on the right can hold one kidney bean, and each big square can hold ten.

You also need a spinner, which you make by cutting out a square piece of oaktag, approximately 10 × 10 centimeters in size. Draw a circle in the middle of the square, and divide the circle into quarters. Then divide each quarter into thirds. You now have a circle divided into twelve segments.

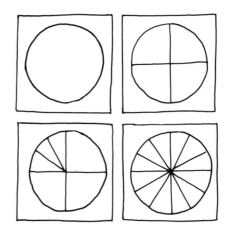

Fill in the segments so the spinner looks like this:

Next you need an arrow to spin. Cut this out of oaktag, too. Attach the arrow to the center of the circle with a paper fastener. Wiggle the arrow until it moves freely, and your spinner is complete.

Finally, you need a bowl filled with at least one hundred dried kidney beans.

The object of the game is to be the first to get fifty beans on the board. The five large boxes on the left side of the board hold ten beans each. The nine small boxes on the right hold single beans. To win, you have to fill all five large boxes with ten beans apiece—that's FIFTY WINS.

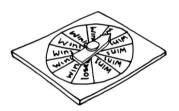

The spinner holds the key to success or failure. Spins end in one of three ways. If you're very lucky, you'll get *win 10.* Then you can take ten beans and fill up a large box on your board. If you're a little lucky, you'll get *win 1.* Then you take a single bean and place it in a small box. When you collect ten single beans you can transfer the whole bunch over to a large box. If you're not lucky at all, you'll get *lose 10.* Then you must remove ten beans from a filled-up large box and return them to the bowl. If you don't have beans in a large box, you have nothing to return. You don't have to return beans that are in the small boxes.

FIFTY WINS is a hair-raising game with many ups and downs that some children find frustrating and others find thrilling. Here's how the game went between me and Hillary, a second-grader who liked this kind of competition.

Hillary started the game by spinning *win 1*. She took a single bean from the bowl and put it on her board. My turn came next. I got *win 1* too. I took a single bean and put it on my board. Tie score, so far.

**Hillary**          **Peggy**

Hillary went again. She got another *win 1* and added a second bean to her board. On my next spin, I got *win 10*. I took ten beans from the bowl and put them in one of my large boxes.

**Hillary**          **Peggy**

Then Hillary lucked out: she spun *win 10.* I was not so fortunate:
I got *lose 10.* I had to take ten beans *off* my board.

**Hillary**          **Peggy**

Up and down, that's FIFTY WINS.
After many more spins, our game boards showed:

**Hillary**          **Peggy**

Things were looking pretty good for me, but in FIFTY WINS
things aren't always what they seem. It was Hillary's turn. She
took her spin, got *win 1,* and added a new bean to the nine single
beans already on her board. That gave her ten single beans. In

FIFTY WINS when you get ten single beans, you group them together and place them in one of the large boxes. Why? Because in our number system, we group tens. That's how Hillary went from having two tens and nine singles to having three tens and no singles. Now it was my turn. I got *win 1*, making my score forty-six.

**Hillary**          **Peggy**

Next, Hillary got *win 10*. All I needed for victory was a *win 10*. Sadness and sorrow, I got *win 1*.

**Hillary**          **Peggy**

It was Hillary's turn again. She gave the arrow a strong whack.
It spun around and landed on *win 10*—we had a winner!

**Hillary's Winning Board**

# BAG OF CHIPS

**GRADES**

two and three

**MATERIALS**

paper
pencil
small paper bag
nine blue, nine red, and
nine green poker chips
(you can substitute multi-colored
bingo markers or small construction-
paper squares)

**G**o into any second-grade classroom and ask the children to write down the number sixteen. I bet at least one child will write 61. Then ask the class to write the number one hundred forty-six. I'm sure several children will write 100406. Next enter a sixth-grade class and dictate the number six million, five thousand, four hundred twenty-three. When you look at the children's papers, you're bound to see a few of them marked 65423. Why do so many children confuse 16 and 61? Why do so many children, even in sixth grade, have a hard time writing large numbers? These errors, from sixteen to six million, happen because children don't understand the mathematical notion of place value. They don't understand that in our number system there's a ones place, a tens place, a millions place, and that 6 in the ones place means six, 6 in the tens place means sixty, and 6 in the millions place means six million.

How to help children understand this concept? With Jamie, a second-grader who needed a little extra help, I played a game called BAG OF CHIPS. Jamie thought BAG OF CHIPS was nifty because we used poker chips to create numbers, and whoever created the biggest number won the game.

Before we started to play, I needed to make two paper scorecards, one for Jamie with his name and two dashes on it, another for me with my name and two dashes. I called the dash on the left the tens dash and the dash on the right the ones dash.

Next, I put nine blue poker chips and nine red poker chips in a paper bag. We were ready to play. I put my hand in the bag and—without looking—took out ten chips. When I finished, I counted my blue chips and wrote that number in my tens place. Then I counted my red chips and wrote that number in my ones place. The first time we played, I picked six blues, and four reds, and got a score of 64.

As soon as I finished recording my score, I put the chips back in the bag so Jamie could take a turn. He picked, he counted chips, and he wrote his score. Poor Jamie, he only got three blues and seven reds.

My number was bigger, so I won. Jamie wasn't one to moan and groan about bad luck, however. Instead he was eager to take me on for a second round.

It didn't take many rounds before Jamie realized that blue chips were more valuable than red chips. He also realized that a four in the tens place is more valuable than that same four in the ones place. Playing this simple poker-chip game helped Jamie understand one of the most difficult and important topics in elementary-school mathematics.

After Jamie became expert at this version of BAG OF CHIPS, I upped the ante. This time I drew three dashes on the paper:

I called the left-hand dash the hundreds place, the middle dash the tens place, and the right-hand dash the ones place. Then I added nine green chips to the bag. I told Jamie that now he could take fifteen chips in all. When he finished picking chips, Jamie wrote the number of green chips he'd accumulated in the hundreds place, the number of blue chips in the tens place, and the number of red chips in the ones place.

This version of the game was exciting, because startling things could occur. For instance, I might pick seven blue chips and eight red chips from the bag, and all Jamie needed to beat me was one green chip. One little measly green chip and all my reds and blues would come to naught.

That was as far as Jamie and I went with BAG OF CHIPS. We could have gone on into the thousands, but Jamie wasn't familiar with such large numbers. Children only advance, after all, one chip at a time.

**T**HROW A NUMBER is a variation of BAG OF CHIPS. Make paper scorecards for you and the child with two dashes side by side, like this:

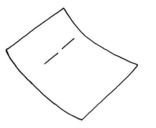

**GRADES**
two and three

**MATERIALS**
paper
pencil
two dice

The dash on the left represents the tens place. The dash on the right is the ones place. You play by rolling a pair of dice and using the numbers to fill in the dashes. It's up to each player to decide which number goes in the tens place and which goes in the ones place. When you toss a four and a two, you can arrange the numbers like this:               or like this:

The winner is the person with the highest number, so 42 is the better choice.

Your score is set. Now it's your child's chance to out-roll you. He shakes, throws, and gets a five and a one. He can arrange the numbers like this:               or like this:

If your child knows his numbers, he'll pick 51 and win the game. (If the child doesn't know which is larger, 15 or 51, then hold off on this game.)

Did you and your child have fun in round one? Then skiddoo to round two. Or challenge your child to the best three out of five rounds, or five out of seven. You might make the grand champion the player who wins ten rounds first.

Once your child understands how to play, try introducing a more advanced version. In advanced THROW A NUMBER you draw two dashes side by side on your scorecards, just as in the beginner's version. But now you throw one die two different times. Let's say that the first time you throw you get a three. You can put this three in the ones place                    or in the tens place.

Your decision involves some thought. You want a big number in your tens place. You know there are six numbers on the die: 1, 2, 3, 4, 5, and 6. You know you'll get to throw the die again. You must think about your next throw. Will you get a bigger number than three? If you believe so, you'll want the three in your ones place. On the other hand, if you believe you'll throw a smaller number, you'll want the three in your tens place. You weigh the choices, and put the three in your ones place.

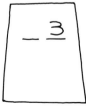

Next, your child throws the die. He gets a two. He must follow the same thinking process you went through. After careful consideration, he puts the two in his ones place.

It's your turn again. You throw a two. Bad luck, the two must go in your tens place.

You still have a chance to win, but only if your child throws a one or a two. Your child shakes the dice. What will he get? Who will win THROW A NUMBER?

As your child develops skill in THROW A NUMBER you can make the game harder by adding a third dash, like this:

Call the dash on the left the hundreds place, the dash in the middle the tens place, and the dash on the right the ones place. In this version of the game, you throw the die three different times. Let's say you get a three on your first throw. You must decide which dash—ones, tens, or hundreds—to place that three on. Try thinking about your choices aloud.

"I don't want to put a three in my hundreds place. Three is too small a number for that! Maybe three should go in my ones place. But then if I throw a smaller number next time, like a one or a two, I'll have a lousy number in my tens place. I guess I'll put the three in my tens place. I think that's safest."

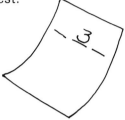

Your child takes the next turn. He throws the die and gets a six. He decides to place the six in his ones place.

You know this is a bad idea, but don't interfere. Let your child make his own choices and his own mistakes. In time he'll find out that the ones place is no place for a six. Children develop number sense quickly when it means winning or losing a game.

You throw again and get a five. Evaluate the situation, and share your strategy with your child.

"Should I put this five in my hundreds place or my ones place? Five is a high number. I want a high number in my hundreds place. I think I'll put the five there."

Your child throws. He gets a two, and plops it in his tens place.

You throw the die and get a one. You fill in your last dash.

Your child throws the die and gets a four. He fills in his last dash.

You won this round, 531 versus 426.

Here's another way to vary THROW A NUMBER. Change the goal of the game. Instead of trying to make the biggest number you can, try to make the smallest number. In this version 426 beats 531. In order to win, you and your child have to reverse game strategy. Your child can only do this if he really understands the game and the mathematical idea behind the game. Don't be surprised if the child makes silly mistakes when trying for the smallest number. Be patient. As long as you're patient, the game will be fun. As long as the game is fun, your child will want to play. Playing is learning.

# THREE POTS

**GRADES**

two and three

**MATERIALS**

three plastic containers
nine dried kidney beans
(or paperclips)
paper
pencil
scissors
masking tape

**T**HREE POTS is a variation of BAG OF CHIPS and THROW A NUMBER. Why another variation on the same game? Children, like adults, learn difficult concepts more thoroughly when the ideas are presented many different times and in different ways. The idea that a single symbol—for instance, the digit 3—represents different amounts depending on its placement in a given number is an extremely difficult concept. Children need to encounter it in many different ways. According to Joey, THREE POTS was the very best way.

Joey was an active second-grader. He was talented in math, but he found it hard to sit still and finish the exercises in his workbook. THREE POTS gave him a chance to learn math while playing an action-packed game.

Before the game began, I collected three plastic containers. The pint containers grocery stores use to pack cole slaw are fine for this game. On the first container, I wrote the word ONES. On the second container, I wrote the word TENS. On the third container, I wrote the word HUNDREDS. When we were ready to play, I placed these containers in ranking order on the floor. Then I placed a strip of masking tape on the floor, two to three feet away from the first container.

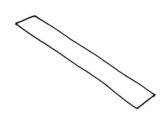

Joey and I took turns sitting behind this line throwing dried kidney beans at the containers. On Joey's turn, he pitched nine beans, one bean at a time. Meanwhile, I sat next to the containers and played catcher. When Joey failed to land a bean, I handed it back to him. Joey's turn didn't end until all nine beans were settled in one or another container.

When all beans were in place, we tallied up how many hundreds, tens, and ones he'd accumulated. We wrote this on a score sheet. The score sheet was a piece of paper with three dashes on it.

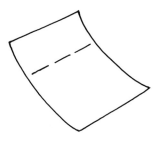

We called the dash on the left the hundreds place. That's where we wrote the number of beans that Joey managed to get in the hundreds container. We called the dash in the middle the tens place. That's where we wrote the number of beans Joey landed in the tens container. We called the dash on the right the ones place, and there we wrote the number of beans that landed in the ones container.

If Joey got five beans in the hundreds container, two beans in the tens container, and two beans in the ones container, his score was 522. If Joey got eight beans in the hundreds container and one bean in the ones container, his score was 801.

After Joey finished his turn and recorded his score, it was time for us to switch places. I sat behind the masking tape and threw beans at the containers. Joey sat beside the containers and retrieved my misfired beans. When all nine beans were settled in containers, we tallied my score. Then we compared scores. Whoever had the highest number was the winner.

Over the course of several weeks, we played THREE POTS a lot. Joey's comments as we played revealed his growing awareness of how numbers work. Once, when I tossed a bean and it landed in the ones container, Joey chided, "Ha, ha, you only got a one. You have to get hundreds if you're going to beat me." Clearly, Joey now understood that a bean is worth more in the hundreds container than in the ones container. It was the same bean, but place (its container) changed its value.

During another game, Joey calculated my score after I'd thrown five beans. "Two in the hundreds, two in the tens, one in the ones," he said. Then he traced these numbers on his hand. "That's two hundred twenty-one." I threw the next bean and it landed in the tens container. "Two hundred thirty-one," he shouted without a moment's hesitation.

How had Joey calculated my new score so quickly? He observed that adding a bean to the tens container did not affect the hundreds or the ones container. Therefore, all he had to do was change the number in the tens place.

Although he wasn't aware of it, Joey had stumbled onto an important mathematical fact. In our number system, it's as easy to add ten to a number as it is to add one. In general, it takes a long time before children understand this. Initially, children add ten by counting ten individual numbers. To add $23 + 10$ they count: 24, 25, 26, 27, 28, 29, 30, 31, 32, 33. Joey's lack of hesitation when adding ten to my score showed he was beginning to make a change in his way of adding. He was beginning to think: "$23 + 10$ . . . all I need to do is count one more in the tens place, and that's 33." THREE POTS gets three hurrahs for helping Joey transform his mathematical thinking.

**G**ail was in the fourth grade, and not doing too well in math. I asked her to solve an addition problem and to talk out loud as she did it, so I could hear the steps she went through. The problem was:

This is what Gail said. "First I add eight and five. That's thirteen. So I write the one and carry the three.

"Then I add five and seven. That's twelve. So I write the two and carry the one.

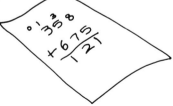

"Then I add three and six and one. That's ten. So I write the one and carry the zero.

"Zero is zero, so I can forget it. Then I'm done."

# GROUP TEN

## GRADES
two and three

## MATERIALS
paper
pen
a deck of cards with picture cards, Jokers, and Tens removed
some yellow, blue, and red construction paper
optional: thirty pennies, thirty dimes, and one dollar

What explained this mixed-up procedure? I asked her why, when she added eight plus five and got thirteen, she wrote the one and carried the three.

Gail was taken aback. "Oh, I got it wrong. I'm supposed to write three and carry the one," she said. Why did she change her mind? She shrugged her shoulders and replied, "You write one number and carry another. That's all I know."

Now I understood. Gail had tried to memorize the rules for adding, but she didn't know the reasons for the rules. Because her actions had no meaning, she got confused and forgot what to do. GROUP TEN gives meaning to the rules. That's why it was a good game to play with Gail.

We needed two paper boards, one for me and one for Gail. Each board looked like the one on the left. We also needed a deck of playing cards with Tens, picture cards and Jokers removed. Finally, we needed a set of thirty yellow markers, thirty blue markers, and one red marker. I cut up small squares of construction paper to make these markers. I could have saved myself the trouble, however, by using poker chips.

| red | blue | yellow |
|-----|------|--------|
|     |      |        |

The idea is to capture the one and only red marker. Players take turns drawing from the deck. The card tells you how many yellow markers to put on your game board. If the card is a Three, you put three yellow markers. If an Eight, you put eight yellow markers. The yellow markers go on the section of the board labelled *yellow*. When you finally collect ten yellow markers, you exchange them for one blue marker. You put the blue marker on the section of the board labelled *blue*. The first player to collect ten blue markers exchanges them for one red marker and wins the game. Before starting an actual game with Gail, I set up some what-if situations to help her understand exactly how GROUP TEN works.

"Gail, what if on your first move you draw a Four from the deck. What will your board look like?"

Gail thought a moment, then she took four yellow markers and placed them on the game board.

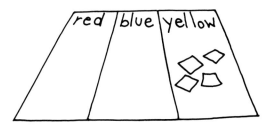

"That's right." I said. "What if your next card is an Eight? What will your board look like then?"

Gail carefully counted out eight yellow markers and put them on the board.

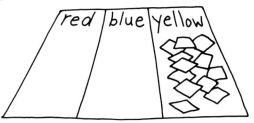

"That's OK," I said. "But remember, whenever you have ten yellow markers you can group them together and exchange them for a blue marker. Do you think you have enough yellow markers for an exchange?"

Gail counted out ten yellow markers, grouped them together, and exchanged them for a blue.

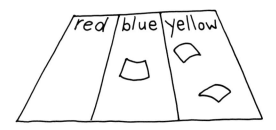

"All right now, what if your next card is a Two? What will your board look like?" Gail picked up two yellow markers and added them to her board.

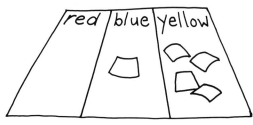

"Now, what if you get a Six?" Gail counted six yellow markers and put them on the board. Then she counted all the yellow markers. She had enough to form a group of ten.

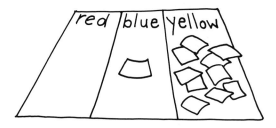

She took the ten yellow markers, exchanged them for a blue, and her board looked like this:

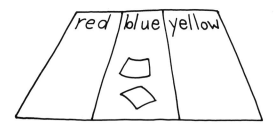

"That's the idea. Now let me set up the board differently. Let's say you have nine blues and six yellows." I put nine blue markers and six yellow markers on the board.

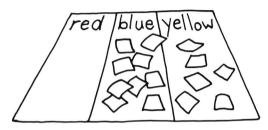

"What if the next playing card you pick is a Seven? What will happen to your board?" Gail took seven yellow markers and put them on the board.

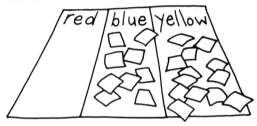

Then she grouped ten yellows together and exchanged them for a blue.

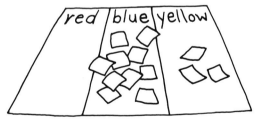

"Count those blue markers, Gail. How many do you have?" Gail counted. She had ten. "Do you know what that means? You can group ten blues together and exchange them for a red. If this had been a real game, you'd be the winner!"

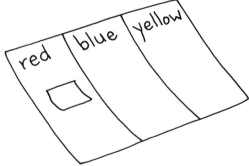

After this demonstration, Gail and I played a real game. In fact, we played many real games over the next few months. How did GROUP TEN help Gail? The game is a physical model demonstrating the rules for addition. Playing it gave Gail a bit of practice working with the number system. It took the abstract concepts of mathematics and made them colorful, visible, parts of an enjoyable experience. And with this practice, Gail came to understand why she had to "write the three and carry the one." Once she understood the meaning behind the rules in addition, math became less baffling and more fun.

With a little variation, GROUP TEN also helped Gail with subtraction. What was the variation? We played the game in reverse. When the game began in the reverse version, I had one red marker on my board and Gail had one red marker on hers.

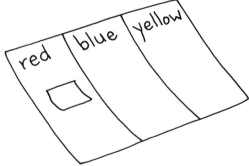

The goal of reverse GROUP TEN was to empty our boards of all markers. Again, we took turns drawing cards from the playing deck. This time, however, the cards told us how many yellow markers to *remove* from the game board. Let's say my first card was a Four. I would have to remove four yellow markers. But I don't have any yellow markers. If I had a blue marker, I could exchange it for ten yellow markers. But I don't have any blue markers. What can I do? I take the red marker and exchange it for a group of ten blues.

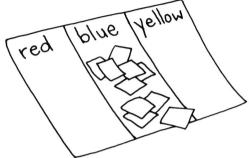

Then I take one blue marker and exchange it for a group of ten yellow markers.

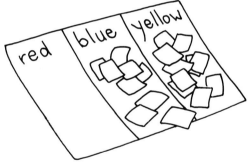

Now I can remove four yellow markers.

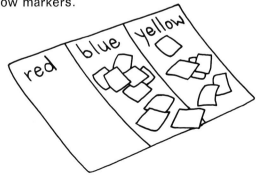

That's how reverse GROUP TEN goes. You keep exchanging blue markers for yellows and removing yellows until the markers are all gone. The first person to empty his board wins.

If you think your child would enjoy a longer game, try starting off with two or even three red markers. Then play and play until there is nothing left. If your child would prefer a shorter game, start with five or six blue markers and get to zero from there.

Here's a money-grubbing way to vary GROUP TEN. Instead of playing the game with yellow, blue, and red markers, try playing pennies, dimes, and a dollar. In this variation, the cards tell you how many pennies you can take from a "bank." When you collect ten pennies you exchange them for a dime. When you get ten dimes you exchange them for a dollar. Whoever cashes in for the dollar first, wins the game. With a nod to Harry Truman, I call this variation THE BUCK STOPS HERE.

# chapter 9

# The Bigger the Better

Children need lots of practice when learning to add and subtract large numbers, especially when they have to "carry" or "borrow" or "exchange" (or whatever word your child uses.) In school, most of this practice takes the form of workbook drills. Children solve problem after problem and eventually get the idea, or at least get into the habit of doing what they're supposed to do. Practice is effective—and boring. But there is, fortunately, an alternative, which is to play the right kind of game. Games are not boring, yet they still offer a lot of drill.

You should be careful about when to play the games in this chapter. You don't want to get into a situation where you yourself end up teaching your child how to exchange, carry, or borrow, the procedures of adding and subtracting big numbers. Possibly you could teach these procedures if you wanted to, but then your child is likely to encounter the same procedures being taught at school in a different style, and the result will be confusion. It's best, therefore, to wait until your child already knows how to add and subtract with big numbers, which will probably be some time during third grade. Then bring out the games to provide needed practice in doing what the child already knows in principle how to do. It's a matter of helping the child polish his act.

BIGGEST AND SMALLEST is a dice game that gets children to add or subtract equations like these:

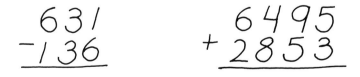

with nary a complaint. NUMBER TRAILS uses a calculator to keep high-tech children interested in good low-tech paper-and-pencil arithmetic. NUMBER ESP teaches your child astounding number tricks that, incidentally, encourage him to practice addition and subtraction. I've had success with all these games, but not with every child. Tastes differ. Some children are fascinated by NUMBER TRAILS but can't be bothered with NUMBER ESP. Other children can't get enough of NUMBER ESP, but think NUMBER TRAILS is a bore. If you try a game once or twice, and your child makes a sour face, forget it. The activities should be fun, or not done.

**G R A D E**

three

**M A T E R I A L S**

two dice
paper
pencil

*T*ake two dice and roll them on a table. Use the numbers that pop up to make the biggest two-digit number you can. Write this number on a sheet of paper.

Throw the dice again. This time make the smallest two-digit number you can. If this second number happens to be bigger than the first throw the dice again. When you have a smaller number, write it below the first number on your paper.

Subtract the bottom number from the top. The answer is your score. Write it on a separate sheet of paper.

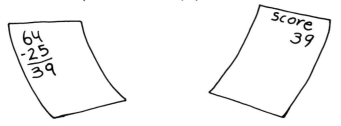

Your child's turn comes next. He throws the dice, makes the biggest number he can, and writes this number on a sheet of paper.

He throws the dice a second time, makes the smallest number he can (he might have to roll again if the second number comes out bigger than the first), and writes it on his paper.

He subtracts. His answer is his score, and he should write it down on another sheet of paper.

The game score stands at 39 to 16. You're ahead, but the game isn't over. You win BIGGEST AND SMALLEST by accumulating a score of one hundred points or more. So far neither you nor your child has reached one hundred. You take another turn with the dice, make a subtraction problem, and subtract again. Now you have more points to add to your score.

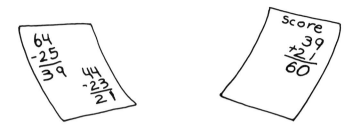

Since you haven't hit one hundred, your child gets another turn. He rolls, makes a subtraction problem, and adds the results to his score. If he breaks one hundred, he wins. If not, you go again. And so on, to someone's triumphal achievement of one hundred or more.

Obviously, in this game the player who goes first has a huge advantage. Clever children usually figure this out and want to go first all the time. You can go along with this, but you don't have to. Instead, you can assert parental standards of fairness by insisting that in BIGGEST AND SMALLEST players take turns going first.

When your child can cope easily with adding and subtracting two-digit numbers, try playing the game with three-digit numbers. To do this, throw three dice and make the biggest number you can.

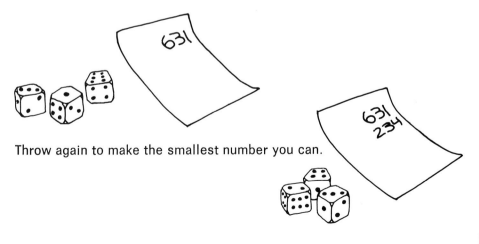

Throw again to make the smallest number you can.

Subtract to find your score.

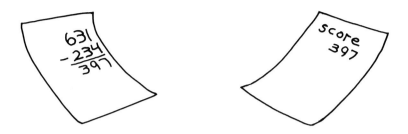

In three-digit BIGGEST AND SMALLEST, the first player to rack up one thousand points is the winner.

Eventually you can play with four dice, make four-digit numbers, and go for a score of ten thousand points.

BIGGEST AND SMALLEST only takes ten or fifteen minutes to play, but in that short amount of time most players solve between six and eight arithmetic problems. Not bad for a bit of extra-curricular mathematics.

**N**UMBER TRAILS is an arithmetic problem that goes "And then . . . and then . . . ," like a good story. Here's an example: take twenty-five and add five. Then add ten. Then subtract four. Then add two, subtract eight, add forty, add three. What have you got? If you follow the trail correctly, you got seventy-three.

When you play NUMBER TRAILS, your child may actually look forward to solving such convoluted arithmetic problems. Strange but true. It's because the child tests his skill against the mathematical abilities of a calculator.

The game works like this: you dictate an arithmetic problem like the one above. You enter the numbers on a calculator while your child works with a pencil and paper. When your problem comes to an end, compare the number on the calculator display with your child's written work. If your child manages to traverse the trail correctly, shout hooray! If he wanders off the trail, commiserate.

Keep your NUMBER TRAILS short and simple when you introduce the game. Eight plus two, plus one, subtract four is a good beginning trail. If your child seems confident with this easy trail, lengthen his travels: eight plus four, plus six, subtract two, plus ten, subtract three. In time you can include larger numbers: one hundred thirty-five plus fifteen, subtract six, add three hundred twenty-four, subtract eighteen, subtract one hundred. Oh, that's a killer.

Undoubtedly, your child will want to take his turn at the calculator, sending you off on a NUMBER TRAIL. Fine, but before he takes command he may need some calculator lessons. Go ahead and explain how the machine works. No need to worry that using a calculator will interfere with your child's mathematical development. Calculators are just another way of dealing with numbers, and the more ways, the better.

**GRADE**

three

**MATERIALS**

paper
pencil
calculator

*The Bigger the Better*

# NUMBER ESP

**H**ere's a number trick I played on Elizabeth, a lively third-grader. Without letting her look, I wrote the number 1,089 on a sheet of paper. I folded the paper tight, handed it to Elizabeth, and told her to tuck it in her pocket. Then I told her to pick any three-digit number so long as each digit was different, and write it on a fresh sheet of paper. Elizabeth chose 579. Next I told her to reverse the digits in her number, and write the new number on her paper.

$$975$$

Now she needed to subtract the smaller from the larger number.

$$
\begin{array}{r}
975 \\
- 579 \\
\hline
396
\end{array}
$$

Next she had to take this answer and add to it those same digits written in reverse.

$$
\begin{array}{r}
396 \\
+ 693 \\
\hline
1,089
\end{array}
$$

Finally I told her to unfold the paper hiding in her pocket.

When Elizabeth saw the matching numbers she let out a screech. "How did you know that?"

For a few bewitching moments I claimed to have magical extra-sensory powers, but eventually I disclosed the truth. You can take any three-digit number, reverse the digits, subtract, reverse the digits of this answer, add these new numbers, and wind up with 1,089. (There is one numerical exception to this pattern, about which, more in a minute.)

I asked Elizabeth if she would like to practice this NUMBER ESP trick and try it out on friends, parents, baby-sitters, and other unsuspecting individuals. She agreed and contentedly spent the next fifteen minutes adding and subtracting until she had the trick down pat.

I warned her about the one exception that crops up. If the first and last digits are consecutive numbers, for instance 374 or 746. you won't get 1,089. You'll get 198. We tried out some problems of this type so Elizabeth could test the results for herself.

What should Elizabeth do if the person she is trying to trick writes one of these exceptional numbers on their paper? I told her to announce a sudden, powerful message from the beyond. She should then grab a fresh sheet of paper and write down 198. Fold this new paper up tight, and hand it over to the person she is fooling. Then she can let the subtraction and addition begin, assured that all will go well.

Elizabeth liked this trick a lot and, over the next few weeks, bragged about how she used 1,089 to bamboozle everyone in her family. So I decided to teach her a new, slightly more complicated number trick.

This trick calls for NUMBER ESP with four digits. I secretly wrote the number 6,174 on a slip of paper. I folded the paper and handed it to Elizabeth. Then I told her to pick any four-digit number with no identical digits (that is, 4,444 won't do). Elizabeth picked 6,930 and wrote this number down on a sheet of paper. I told her to arrange the digits to make the largest possible number.

9,630

then rearrange the digits to make the smallest possible number.

0,369

I told her to subtract the smaller number from the larger.

$$\begin{array}{r} 9,630 \\ -\,0,369 \\ \hline 9,261 \end{array}$$

Next I told her to take the four digits in her answer, and arrange them to form the largest possible number,

$$9,621$$

and the smallest possible number,

$$1,269$$

and subtract again.

$$\begin{array}{r} 9,621 \\ -\,1,269 \\ \hline 8,352 \end{array}$$

Then she had to take the four digits in her answer and repeat the same manipulations. That is, she had to reverse the order of digits and subtract the smaller number from the larger.

$$\begin{array}{r} 8,352 \\ -\,2,358 \\ \hline 6,174 \end{array}$$

Pop! My magic number appeared. I told Elizabeth to unfold the paper I'd given her before she'd done all this calculating. She did and saw the matching number. She was less amazed by this second trick, but equally delighted at the prospect of using it to dazzle friends and relatives.

I pointed out to Elizabeth that the main difference between this four-digit trick and the old three-digit variety is that there's no way to know how many calculations it takes before 6,174 appears. The person you are tricking must keep at the pattern. You must check the answers to each problem. Eventually 6,174 will turn up —and then you grandly unveil the folded paper.

I still had one more number trick up my sleeve. I pulled it out to brighten a rainy day in February. I wrote the number nine on a sheet of paper. I didn't hide the nine; instead, I bragged that if Elizabeth picked any number in the world (except a string of identical digits, like 777,777), I could make her go from that number to the number nine.

Elizabeth agreed to test me out. She picked the number 235,790, and wrote it on a sheet of paper. I told her to rearrange the digits of her chosen number in any way whatsoever.

$$905,237$$

Next, I told her to subtract the smaller number from the larger one.

$$\begin{array}{r} 905,237 \\ - \ 235,790 \\ \hline 669,447 \end{array}$$

I told her to add the digits in her answer.

$$6 + 6 + 9 + 4 + 4 + 7 = 36$$

Finally I told her to add the digits in this answer.

$$3 + 6 = 9$$

What do you know, nine! That's the way it always is. After rearranging, subtracting, and adding, when you reach a single digit, the digit is always nine. Check it out.

Why do these tricks work, you ask? I've put the mathematical analysis in an appendix. The explanations are difficult to follow and demand algebraic thought on your part. If you give up halfway through, just remember, the tricks are fun even when the whys and wherefores remain mysterious.

# PART THREE

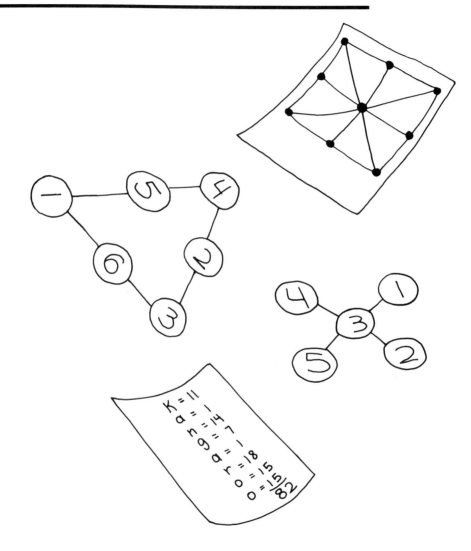

# chapter 10.

# Strategy Games

Consider what a good chess player does: he proceeds in an orderly, logical fashion. He evaluates all the moves available to him at a given moment. He makes changes in his plans if his predictions don't work out. Sometimes he ignores the obvious move in favor of the more obscure. He spices his hard logic with a dash of creativity.

That's exactly how a mathematician deals with a mathematical problem. The orderly, logical approach, the evaluation of options, the changing of plans, the creativity and surprise are no different with digits and symbols than with knights and pawns. Which leads to a question: Does the similarity in thought processes between math and strategy games like chess mean that learning strategy games will help children learn math?

One year I was teaching a combined second- and third-grade class. At the end of the term my principal made an interesting observation. She noticed how enthusiastic the children were at solving tricky math problems. The children in this class didn't seem to get frustrated by difficulty. They seemed actually to enjoy challenges. And she noted, truthfully enough, that my class the preceding year wasn't quite so animated. What changed, she wondered, from one year to the next?

Naturally, the children had changed, I had changed, many things were different. But I think one long-range class project accounted for the main difference. I taught the combined second- and third-grade class to play a new strategy game every month, and sometimes more often. Usually I taught new games on a Monday, and during the week I assigned the new game as part of the children's work, which was a pleasant change from workbooks. For the rest of the month I let the children pick whichever strategy game they wanted to play. The games were a popular choice during free time. Strategy games became a way of life in this class.

I believe strategy games taught those second- and third-graders how to struggle with and enjoy intellectual problems. There's a pleasure in thinking hard, and those children learned to appreciate that.

Chess, I should add, wasn't part of the program. Some children understand the rules, but the actual strategy is pretty much beyond them, leaving aside the occasional Bobby Fischer. Many children, of course, do know checkers, which is a good game. But the children in my class couldn't play checkers day in and day out for a year, so I hunted around for other possibilities. There are, it turns out, a lot of fun strategy games that *are* good for children. There's TAPATAN from the Philippines. The Koreans play a version of Parcheesi called YUT. There's NINE MEN'S MORRIS, which is so ancient that archaeologists have found paintings of it in Egyptian tombs. Another ancient game is KHALA, which is still played today in Africa.

Most first-graders can learn the rules to these games, but they can't really learn the strategy. They rely on luck instead of skill, which means they aren't getting the full benefit in playing. By the time children reach second or third grade, however, their playing begins to bloom. These older children search for good moves, evaluate possibilities, analyze choices. They recognize their power to determine the game's outcome and therefore play intentionally, not impulsively. These children are ready to analyze the problems presented in the games and work hard to come up with solutions. That's why I suggest waiting until your child is in second or third grade before you introduce the analytic pleasures of TAPATAN, YUT, NINE MEN'S MORRIS, and KHALA.

# TAPATAN

**GRADES**

two and three

**MATERIALS**

paper, oaktag, or poster board
pen
index cards in two colors
scissors

**T**APATAN may be regarded as tic-tac-toe's brainy Filipino cousin. It takes five minutes to teach a child the rules. It takes three minutes to play a round. But it's a fine game, and months or years will go by, and your child and even you will still be happily moving the little markers around.

A TAPATAN board looks like this:

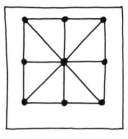

You can draw the board on a piece of paper, or make a more permanent one using oaktag or poster board.

You need six game markers: three for you in one color and three for your child in another color. I usually cut out small pieces of colored index cards, three blue, three yellow, and use them as game tokens. Pennies and nickels would do just as well, or dried lima and kidney beans, black and red checkers, or anything you want.

The idea is to arrange your three markers in a row along any of the TAPATAN board lines. You can get three in a row horizontally, vertically, or diagonally.

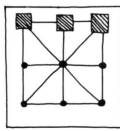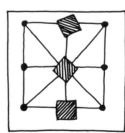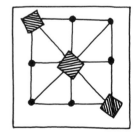

**178**

When the game begins, the board is blank. You and your child take turns placing one marker at a time on any of the TAPATAN board points. When all six pieces are placed, you take turns sliding from point to point along the board lines. You cannot jump over another marker. You cannot share a point with any other marker. You take turns sliding pieces until one player gets three in a row, or until you declare a stalemate.

Simplicity—that's the beauty of TAPATAN. There are only six pieces and nine places to put them. That limits the choices for any single move. Even young children, with adult guidance, can weigh options and plot strategy in a game like this.

I have two techniques to help children develop strategic thinking while playing TAPATAN. First, when it's my turn, I talk about possible moves.

"Unless I think quick, you'll get three yellows in a diagonal row on your next move. What can I do to block you? I'll move one of my pieces to the middle dot. Which piece should I use to block? I can move the piece on the left. That stops you, but it doesn't help me. I can move the piece on the bottom right, but that doesn't help me either. I can move the center piece on the bottom. Now that sets me up to get three in a blue diagonal row, and I don't think there's a thing you can do to stop me. That's what I call strategy."

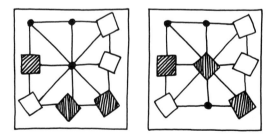

I also ask children to evaluate their options before moving. I don't do this every single time a child takes a turn. That would slow the game to the point of boredom.

When the moment seems right, I say, "There is a fabulous move on the board for you. Look at each of your pieces. Where can each one move? Can you find a move that stops me from getting three in a row and helps you get three in a row?"

If I'm patient and encouraging, sooner or later children begin to analyze their options. They begin to think first and move second. They devise plans, carry them out, and evaluate the results— which happens to be the kind of thinking used by mathematicians.

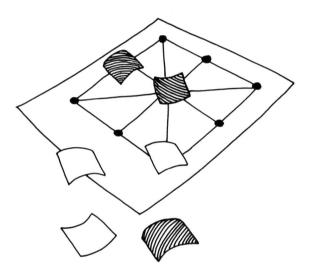

**Y**UT is Parcheesi, Korean-style.

Here's how to play, according to my Korean neighbors in New York, with a few pedagogical modifications of my own.

You need a board, which you can make with paper, oaktag, or poster board.

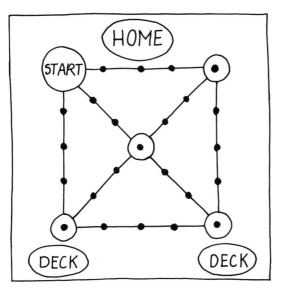

**YUT**

**GRADES**
two and three

**MATERIALS**
paper, oaktag, or poster board
pen
index cards in two colors
scissors
one playing die

Each player needs four game tokens. You can cut out small pieces of index cards—four in one color and four in another—and use them as markers. Or use four pennies and four nickels, or poker chips, or any old thing.

You begin with your four tokens in one of the *Deck* circles, and your child with four tokens in the other. Turn by turn, you and your child take tokens from your *Deck* circles, put them on *Start*, and race them around the board until they reach *Start* again. Then they can go *Home*. The first player to get all four tokens *Home* wins.

The racing goes like this. When it's your turn, you roll the die. The number you get is the number of spaces you can move your

token. The moving always goes counterclockwise around the out-
side lines of the board—

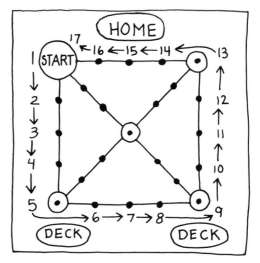

unless you land on one of the bottom corner points. A token that
lands on the bottom left-hand corner point can take a diagonal
shortcut across the square to the top right-hand corner.

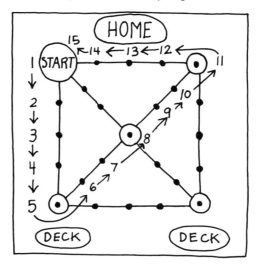

Better yet, if while taking the diagonal shortcut the token lands on the middle dot, it can turn and head directly back to *Start.*

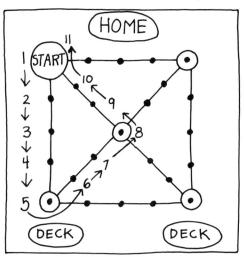

If a token passes over the bottom left-hand corner, it might still land on the bottom right-hand corner point. And if it does, it can take that diagonal shortcut across the square and zip back to *Start.*

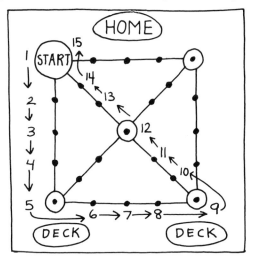

There are three other rules. Each time you roll the die, you can pick which of your tokens to move, or take a new token from your *Deck*, put it on *Start*, and begin its march around the board. If you land on your opponent's token, the opponent's token goes back to his *Deck*. If you land on your own token, you can piggyback the two tokens (or however many you have on that point) and move them as if they were a single token. Piggybacking doubles the speed with which you can get the tokens home, but also doubles the damage if the other player lands on that point.

YUT demands thought. You roll the die: let's say a five comes up. Should you add a token to the board and move it five spaces, or move a token that is already in play? Should you piggyback your tokens or let them travel separately? Should you *not* move a certain token, hoping to get the right number for a shortcut on your next turn? Should you get a token safely back to *Start* or take an opportunity to bounce your opponent back to his *Deck*?

If you chatter about your game plan aloud, your child may learn from your strategic thinking. Occasionally, after your child throws his die, ask him to evaluate his options. Point to his pieces, one at a time, and ask, "Where will you land if you move this piece? Where will you be if you move that one?" After helping your child determine all the possible moves he can make, let him decide which is best.

I like YUT. Millions of Koreans like YUT. But the acid test is, do children like YUT? To find out, I played this game with a group of second- and third-graders that included a lively girl named Carmen. Carmen noted YUT's similarity to Parcheesi. I asked which game she preferred. She answered YUT. But why? She thought for a few moments and explained, "If you're going on a trip and you have a lot to pack, you can still take this game because it's just a piece of paper. You can't take Parcheesi because it's too big. So I like this one better because you can play it anywhere."

*T*here are some mysterious lines in *A Midsummer Night's Dream* where Shakespeare says:

> The nine men's morris is filled up with mud,
> And the quaint mazes in the wanton green
> For lack of tread are undistinguishable.

These lines refer to a game that is very suitable to children, and very good for teaching strategic thinking. The game is called NINE MEN'S MORRIS. It's best if you are a zillionaire and have a large country estate with a green and a maze made of bushes to serve as your playing field. But perhaps you are away from your estate for the moment—in which case you can make a board out of paper, poster board, or oaktag. The board looks like this:

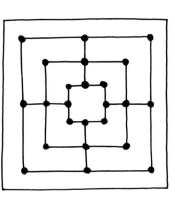

You also need eighteen game tokens—nine in one color and nine in another. I use cut-up bits of colored index cards for tokens, but you can use whatever you have handy at home.

The game begins with you and your child taking turns each placing one token on the board at a time, until all the tokens are on the board. You can place your tokens on any one of the board's twenty-four points. Once all the pieces are in place, you and your child take turns sliding your tokens along the board lines. You cannot jump over pieces and you cannot share a point with another token.

**GRADES**

two and three

**MATERIALS**

paper, poster board, or oaktag

pen

index cards in two colors

Your first strategic goal is offense. You want to get three tokens in a row somewhere on the game board. Your three in a row can be on any horizontal or vertical line. Each three in a row you get is called a mill.

Black has a mill on this board. Can you find it?

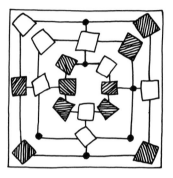

White has a mill on this board. Can you see it?

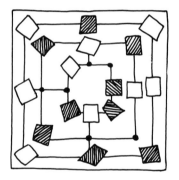

When you get a mill, you can remove one of your opponent's game tokens from the playing board. You can take any of your opponent's pieces, except those your opponent has already arranged in a mill of his own, and even those can be taken away when no others are left. Another exception is that you have to wait until all the pieces have been put on the board before you start forming mills and taking pieces away.

Here's what black did after getting his mill:

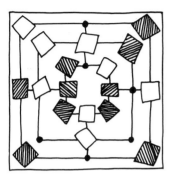

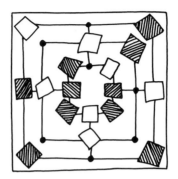

**Before**          **After**

Here's what white did:

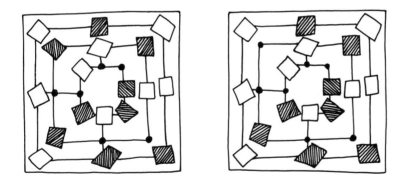

**Before**          **After**

Your second strategic goal is defense. You want to do everything you can to prevent your opponent from forming mills. That way you save your own pieces.

Look at this board—right now it's black's move, and he would be well advised to stop white from making a mill.

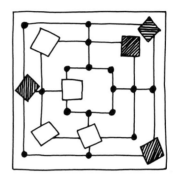

There are two ways to win NINE MEN'S MORRIS. You can remove seven of your opponent's pieces, leaving him with just two tokens on the board, or you can trap your opponent so he can't make a move. Study the board on the left to see how white won by capturing pieces.

On the right, six blacks won by trapping seven whites:

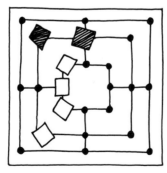

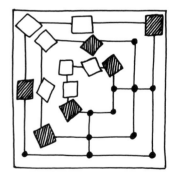

**White wins**                     **Black wins**

There is one tricky NINE MEN'S MORRIS maneuver you need to know before you start to play. Let's say black forms a mill here:

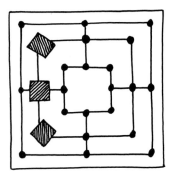

That means black can capture one of white's pieces. What's more, on black's next move he can open this mill (left board). And one move later (right board), he can close the mill again:

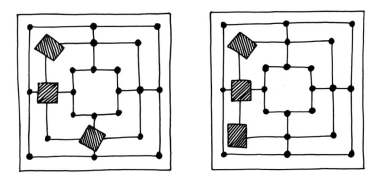

and take another piece away from white. Black can keep opening and closing that mill—which means he can keep taking pieces away from white and will surely win—unless white can figure out how to stop him.

There you have it: NINE MEN'S MORRIS. As you play, your child will probably need to be reminded of the rules. That's OK, just tell him whatever he needs to know. You can also provide a few strategy tips.

"Think very carefully about your next move. Look and see if you can stop me from forming a mill."

"Oooh, I see a great move for you. If you make this move, you'll form a mill and take a piece away from me."

You can do one more thing to help your child appreciate the strategic operations of NINE MEN'S MORRIS, which is to talk aloud about your own moves. Discuss your reasons for sliding a token to this or that spot. Compare the advantages of moving one token rather than another. Allude to plans for forming future mills.

By the way, the game is even older than Shakespeare. Archaeologists have traced it back to ancient Egypt. Some people believe the game was played in the city of Troy. There's evidence the game was known to the Vikings in northern Europe. So when you play NINE MEN'S MORRIS you are joining legendary players, from Cleopatra to Helen of Troy, from Leif Ericson to the bard of Stratford-on-Avon. Who could ask for better company?

# K

HALA is very old. They were playing it in ancient Sumer, 3000 B.C. They play it in Africa today. Soon you'll play it at your house. A good game is a good game.

The first thing you need is a playing board. In Africa, the game is sometimes played using pits dug in the ground, sometimes on beautifully designed wooden game boards. In the United States you can buy commercially made wooden boards. But you don't really need to, since an empty egg carton with the lid cut off and two paper cups will do. You also need game tokens. Africans have used everything from pebbles to diamonds. Are diamonds scarce at your house? Feel free to use dried beans, or paperclips, or anything. You'll need thirty-six of whatever you use.

One warning before you learn (and then teach) KHALA. Of the four strategy games in this book, KHALA is the most difficult for children to learn. It's worth your time and patience to teach the game, however, because KHALA invariably fascinates children and adults alike.

I'll explain how to play by assuming that I am playing a game with you, the reader. We put the egg carton horizontally between us. The twelve egg cups are called pits. (For instruction purposes, I've labelled the pits one through twelve. You don't need number labels for real games, however.) I have six pits on my side of the board, numbered one to six. You have six pits on your side, numbered seven to twelve. The two paper cups go on either end of the egg carton. These are our home pits, and they are special. Your home pit is on your right. My home pit is on my right. Finally, we put three lima beans in each egg carton pit, but not in the home pits. Now the KHALA board should look like this:

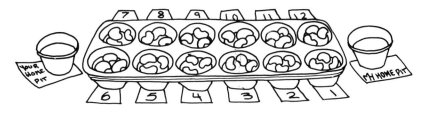

**G R A D E S**
two and three

**M A T E R I A L S**
an egg carton with lid removed
two small paper cups
thirty-six dried lima
or kidney beans

You win by accumulating the most lima beans in your home pit. How to get the beans into the home pits? We take turns moving the beans around the board. On your turn, you take all the beans from any one of your pits, and, moving in a circular fashion around the board, drop one bean at a time into neighboring pits. On my turn I do the same. I'll go first. I take all three beans out of pit number two, I drop a bean in pit number one, a bean in my home pit (a point for me) and then I circle around to your side of the board and drop the last bean in pit number twelve. When I'm done, things stand like this:

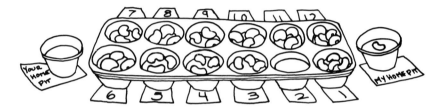

That's the way the game goes. I move the beans in my six pits, you move the beans in your six pits. Beans pile up in pits: sometimes there are enough beans to circle all the way around the board in a single move, one bean per pit. You never drop a bean in the other player's home pit, however. You just skip over it.

It's your turn: you empty pit number twelve. You take all four beans from that pit and, moving in the same circular direction as I did, you drop a bean in pit eleven, a bean in pit ten, a bean in pit nine, and a bean in pit eight. When you're finished the board looks like this:

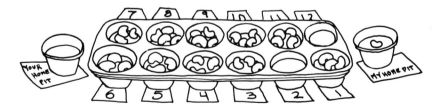

My turn again. I'm going to empty all my beans in pit one. I drop a bean in my home pit and then a bean in pit twelve, a bean in eleven, and the last bean in pit ten.

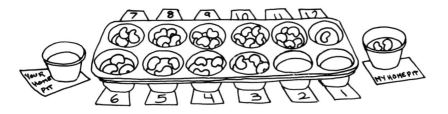

That gives me two home pit beans versus zero for you. It's your turn now, and time to tell you a special KHALA move. If you can arrange for your last bean to land in your home pit, you get an extra turn. Right now, by moving the five beans in pit eleven, you drop off beans in pits ten, nine, eight, seven, and your last bean lands in your home pit. All right, that means you can take another turn.

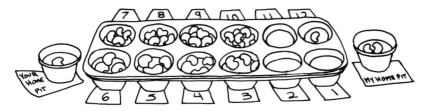

For your extra turn, you move the five beans in your pit nine. Now the board looks like this:

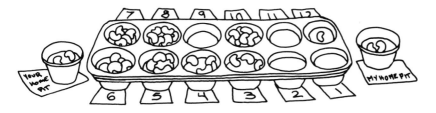

My turn, and I have some good moves. First off, I move the beans in pit three. My last bean lands in my home pit and I get an extra turn.

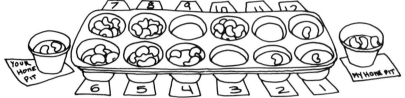

Now I move the single bean in pit one. It lands in my home pit —I get *another* extra turn.

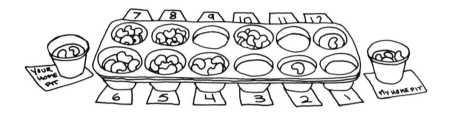

But before I take my second extra turn, I must tell you about another special move. If my last bean lands in an empty pit on my side of the playing board, I can steal some of your beans and place them in my home pit. Right now, if I move the beans in pit four, my last bean lands in pit one, which is on my side of the board and is empty. That allows me to steal whatever beans lie directly opposite this empty pit. Directly opposite is your pit twelve. A single bean sits in pit twelve, and it's mine to steal. My move is over. It's your turn, and here's the board:

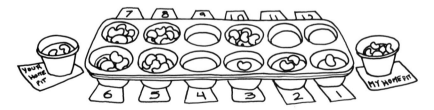

You decide to move the six beans in pit ten.

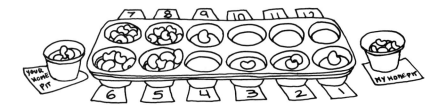

It's my turn. I'm going to move the two beans in pit two. The last bean lands in my home pit, giving me an extra turn.

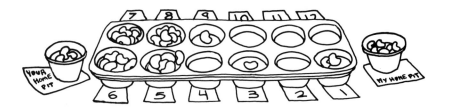

For my extra turn, I could move the single bean in pit three. It would land in my empty pit two. Oh, if you only had ten beans in pit eleven, I could steal them all. Too bad you don't have any beans In pit eleven, so there's not much point in that move. Instead, I think I'll move the five beans in pit six.

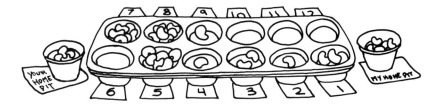

The game ends the first time a player can't go because there are no beans on his side of the board. Then you count up the beans in each player's home pit, and whoever has most beans wins.

Now that you're familiar with the game, you can introduce KHALA to your child. You'll find that the more often you play, the more KHALA strategy you'll develop. This is a real think-it-out game and you'll soon see there's a lot to think about. As you learn clever KHALA moves, why not share your wisdom? For starters, here's a KHALA setup children enjoy mastering. You can use this maneuver whenever you have a single bean in pit seven and two beans in pit eight.

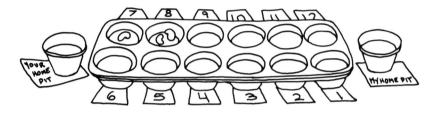

Begin by emptying pit seven and getting an extra turn.

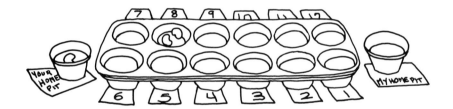

Then empty pit eight and get another extra move.

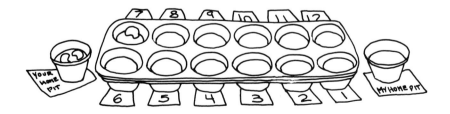

Finally, re-empty pit seven and get one more extra move.

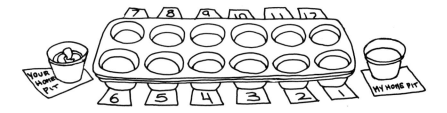

With this setup you just got three moves in a row and three beans safely laid away in your home pit.

Your child may catch on to KHALA right away, and that would be fine. It will also be fine if he has a hard time figuring out the game. If he does have a hard time, you can make things easier. Play eight-pit KHALA instead of the twelve-pit game. To do this, simply cut down your egg carton to make four pits on either side instead of six, and play the game exactly the same way. With fewer pits and fewer beans, there are fewer strategic options for your child to think about. This gives the child more control over the game. Once he masters eight-pit KHALA, cut a new egg carton board with ten pits—five on each side. When your child masters ten-pit KHALA, go to the regular twelve-pit game.

What does KHALA teach? Orderly thinking and strategic calculation. That's not what a child gets from watching TV.

# chapter 11.

# It's a Puzzle

Mathematicians are patient, persistent, curious, logical and systematic. How can you develop these qualities in your child? Playing the strategy games of the last chapter is one way. Playing the puzzle games in this chapter is another.

There are three of these games. When your child plays NUMBER BUBBLES and WHAT'S IT WORTH? he'll add and subtract until he comes up with the right number to fill a bubble or beat you in a number-the-words contest. The necessary arithmetic is simple, but the logical manipulation of numbers will demand your child's full attention. When your child plays COLORED BOXES he won't deal with numbers at all. Instead he will have to use logic and systematic thought to work up patterns with colors.

First-graders may stumble on correct solutions for these problems, but correct solutions are only part of the point. Systematic thinking is what you want to encourage, not just right answers. That's why you should hold off on this chapter until your child is at least in second grade, or even fourth or fifth grade. With gentle encouragement, older children can develop sophisticated thinking. Younger children can't do that and will be better off doing what, in fact, they can do.

**J**ason was cranky. We'd been working on some boring worksheets and he was ready for a break. So I decided to present him with a few NUMBER BUBBLE puzzles, which served his purpose by being fun, and my purpose by being mathematical. I took a fresh sheet of paper and wrote the numbers one through four across the top. Then I drew this design:

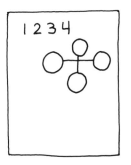

I explained the game: "I have a tricky problem for you to solve, Jason. Do you see these four bubbles? I want you to fill each bubble with one of these four numbers. You have to place the numbers in a special way, however. You must make sure that the numbers going across equal exactly five, and that the numbers going up and down also equal five."

Jason looked a little confused, so I went on with my explanation. "Let's say I put the numbers in the design like this:

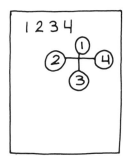

# NUMBER
# BUBBLES

**G R A D E S**

two and three

**M A T E R I A L S**

paper
pencil

Can this be the right arrangement?" Without waiting for Jason, I answered my own question: "It can't be right because the numbers going across equal six, and the numbers going up and down equal four. To solve this problem correctly, the numbers must add up to five across and five up and down. Here, I'll draw a new design on the paper for you."

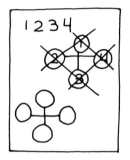

Jason didn't look confused anymore. "I get it. That's so simple for me. My mom says my brain has real noodle power." Sure enough, Jason grabbed a pencil and quickly filled in the correct numbers.

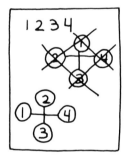

"I agree with your mom. You sure know how to use your noodle. That means I have to make the next problem a little bit harder. This time you have to use the numbers five, six, seven, and eight.

You must arrange them to equal thirteen across and thirteen up and down. Here, I'll draw the bubbles."

This arrangement required a bit more work on Jason's part. First off he filled the bubbles like this:

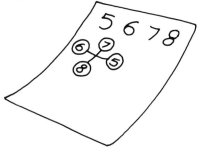

As soon as he added the numbers, however, he recognized his mistake. I drew new bubbles and let him try again. Success.

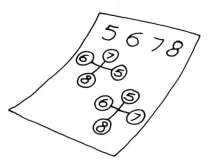

We played NUMBER BUBBLES using four-bubble pictures another few times. I gave him the numbers three, five, seven, and nine, and told him to link them so the bubbles equalled twelve up and down and twelve across. I gave him the numbers ten, twelve, fourteen, and sixteen, and told him to link them so that both directions equal twenty-six.

In the course of solving this problem Jason made a discovery. "Hey, I think I see something. I always add the smallest number and the largest number together and then I add the in-between numbers."

"That's interesting. Let's see if your theory works with these numbers," I said as I made a new bubble design and wrote the numbers ten, twenty, thirty, forty. "Can you make these numbers equal fifty?"

Jason filled in the bubbles correctly with barely a moment's hesitation.

"That's terrific, Jason. And you know what? I think I have to come up with a tougher bubble puzzle for you this time. Would you like to try a five-bubble puzzle?"

Jason agreed to try, and so I drew the following design:

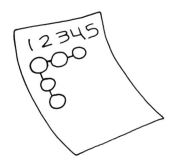

"This time there are five bubbles, and you must fill them in with the numbers one through five. You have to fill the bubbles so that the numbers going across add up to eight and the numbers going down equal eight."

Jason started working across. He filled in the first three numbers.

"That's eight," he said with satisfaction. Then he filled in the numbers going down.

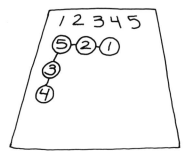

"Wait, that's not eight. What happened?"

"Your numbers going across are OK, but the numbers going down are too big. Can you think of a way to keep the same numbers going across, but switch them around so that you have smaller numbers going down?"

Jason thought this over, "Well, I guess I could move the five and put the two there instead."

"Good idea, why don't you try it," I said as I drew another bubble picture.

"That's still not right. I have to try something new," Jason said as he added his results. "Maybe I should move the number one."

"Fine idea," I said. "Just give me a second to draw new bubbles."

As soon as I finished drawing, Jason filled in the new bubbles. He added, then shouted, "I got it! I got it!"

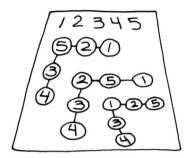

"Wow, Jason, that's just amazing. You are so good at this game, I'll have to come up with an even tougher bubble. How about this? Using the exact numbers and the same design, arrange things so that the numbers equal nine across and nine up and down."

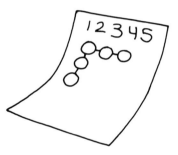

After a few minutes of concentrated effort, and one false start, Jason solved this problem.

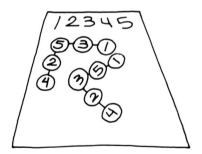

So I gave him another. I told him to use the same numbers and the same design, only now the numbers would have to add up to ten.

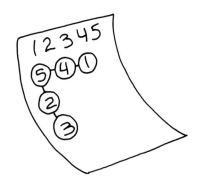

"Boy," said Jason when he finished. "You really have to use your noodle for this game!"

That was enough for one day. In the days and weeks ahead, however, I made frequent use of this mind-stretching activity. I progressively made the problems more difficult. Sometimes Jason discovered patterns that helped him to quickly fill in the bubbles. Other times he tried one number combination after another until he figured out the solution. Here are some of the designs and numbers I called on Jason to arrange. You might want to try them out with your child, for instance, while waiting for a pizza delivery. If you're really into the game, continue playing while slurping that cheese and tomato.

I've listed the puzzles in order of difficulty. For your benefit, the solutions are included.

## Puzzles

Using the numbers 1, 2, 3, 4, and 5, make the bubbles across and the bubbles going down add up to eight; then try for nine; and then ten.

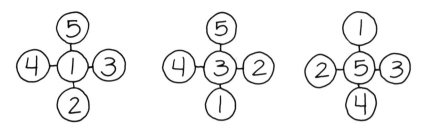

Using the same bubble design and the even numbers 2, 4, 6, 8, 10, make the bubbles equal sixteen, then eighteen, then twenty.

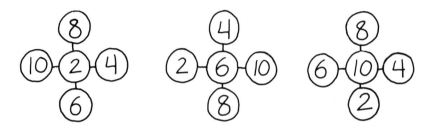

Here's a new design, with seven bubbles. Using the numbers 1, 2, 3, 4, 5, 6, 7, make the numbers going down and going diagonally add up to ten. Then make the bubbles add up to twelve. Then try for fourteen.

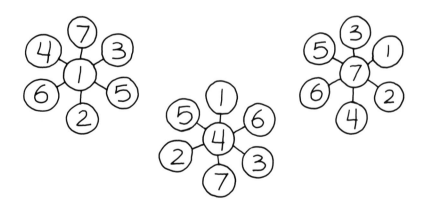

Here's a nine-bubble design. This time, use the numbers 1, 2, 3, 4, 5, 6, 7, 8, 9, and make the bubbles across and down equal twenty-three, then twenty-five, and then twenty-seven.

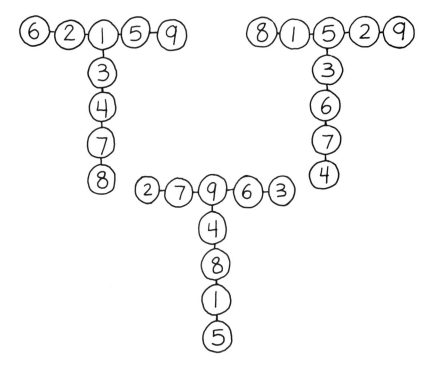

Here's a triangular bubble picture. Use the numbers 1, 2, 3, 4, 5, 6, to make each side of the triangle equal nine, and then twelve.

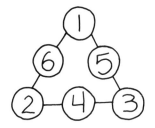 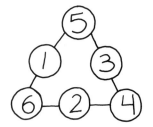

Try the same triangle with the odd numbers 1, 3, 5, 7, 9, 11, and make the sides add up to fifteen, and then twenty-one.

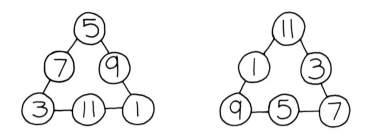

Finally, here's a jumbo triangle. With the numbers 1, 2, 3, 4, 5, 6, 7, 8, 9, make all sides of the triangle equal seventeen, then twenty, and then twenty-three.

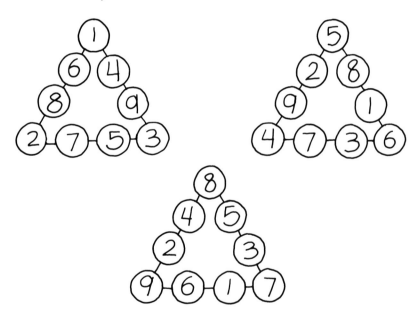

That bubble puzzle is a noodle buster.

**W**HAT'S IT WORTH? is a hard game. It demands a lot of concentration from the children who play it, which means it's not a game for everyone. But for some children WHAT'S IT WORTH? is a great game.

I sat down to play with a third-grader named Yumi. I wrote out the entire alphabet and assigned each letter a number—one through twenty-six.

| a | b | c | d | e | f | g | h | i | j | k | l | m | n | o | p | q | r | s | t | u | v | w | x | y | z |
|---|---|---|---|---|---|---|---|---|---|---|---|---|---|---|---|---|---|---|---|---|---|---|---|---|---|
| 1 | 2 | 3 | 4 | 5 | 6 | 7 | 8 | 9 | 10 | 11 | 12 | 13 | 14 | 15 | 16 | 17 | 18 | 19 | 20 | 21 | 22 | 23 | 24 | 25 | 26 |

"Let's start by finding out how many points are in our names. I'm going to write down the numbers for *p* and *e* and *g* and *g* and *y*. Then I'm going to add up those numbers and know how many points my name is worth."

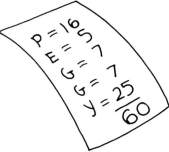

$$P = 16$$
$$E = 5$$
$$G = 7$$
$$G = 7$$
$$Y = 25$$
$$\overline{60}$$

"Looks like I have a sixty-point name. Let's see about you."

Yumi carefully recorded the numbers for *y-u-m-i*, and then she added.

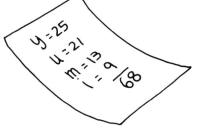

$$y = 25$$
$$u = 21$$
$$m = 13$$
$$i = 9$$
$$\overline{68}$$

**GRADES**

two and three

**MATERIALS**

paper

pencil

optional: calculator

"Sixty-eight!" she proudly announced.

"Great. Now I have another alphabet challenge. Each of us picks an animal. We add up the points in the animal's name and see which beast has the most points. I pick *giraffe*."

I spelled it out, along with the number equivalents for each letter, and totaled them.

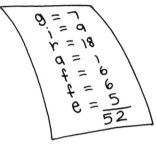

"That's weird," said Yumi. "*Giraffe* has so many letters, but it isn't worth as many points as my name."

Yumi had just made her first intellectual realization in this game: the length of the word does not determine value.

"Good thinking, Yumi," I said. "But if the length of a word doesn't guarantee victory, how can you win this contest?"

Yumi thought. "I'm not exactly sure," she said.

"See if you can come up with an animal to beat *giraffe*. That may help you solve this problem."

"I love lions."

"OK, let's add up the points and see if it beats *giraffe*."

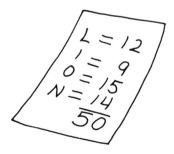

"That's almost as much as *giraffe*. How come? How come *lion* is worth almost as much as *giraffe*?" Yumi asked.

Instead of answering her outright, I said, "Here's a strange one, Yumi. Look at how many points there are in the word *zoo*."

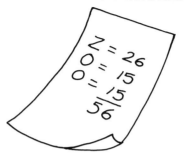

"I think I thought of something. *Zoo* is a little word but it has a *z*, and *z* is worth a lot of points. *Giraffe* has lots of letters but the letters aren't worth so many points."

Yumi had made a second intellectual realization: the value of each letter in a word means more than how long the word is.

"Well, then," said Yumi, "I want my animal to be *zebra*."

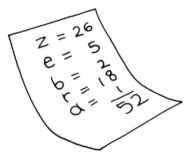

"I thought I would win with *zebra*," Yumi said. "Instead it's just a tie."

"Zebra does have a *z*," I said. "and the *r* is worth eighteen points, which is pretty good. But the rest of the letters *e*, *b*, and *a* aren't worth much. I'm curious about *kangaroo*. *Kangaroo* doesn't have a *z* but it has some letters that are worth lots of points, and it's a big word, too."

"OK, let's try it," said Yumi.

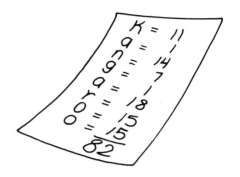

"Wow, eighty-two points," said Yumi after adding.

I was pretty sure Yumi had the intellectual framework to see how to win: you need to balance the length of a word against the point value of the individual letters. So I proposed a new contest. "Let's see who comes up with the most foods that are worth less than fifty points."

This contest was tricky. To win you need to try out lots of different foods. You use logic to think of short words with letters low in the alphabet. You would never even experiment with *Zwieback*—a quick estimation of the numerical values would prove it a loser. *Ice cream*, however, is worth forty-seven points, so it's a go. Given the result with *ice cream*, would you give *egg cream* a try?

It's not addition that makes this game so valuable. It's the numerical estimation and the logic that makes you try one word and let the information you gain lead you to try other words. That's why, when Yumi asked if we could use a calculator to make the addition part of the contest move a little faster, I was quick to agree. It is also why, since Yumi liked these contests so much, I came up with a new WHAT'S IT WORTH? every few weeks. Here are a few versions Yumi and I played that you might want to try at home.

Find the most words worth less than twenty-five points.
Find a word worth exactly seventy-five points.
Be the first to find three names worth between twenty-five and fifty points.
Find a word worth more than one hundred points.
In ten minutes, find the most words worth between fifty and one hundred points.

Yumi got a kick out of these competitions. Perhaps you and your child will, too. If so, WHAT'S IT WORTH? may be a favorite game in your house. Many people, however, who are as bright and studious as can be, hate this kind of activity. If that's you, or if that's your child, forget about WHAT'S IT WORTH? *Games for Math* has fifty-five other games you can play instead.

# COLORED
# BOXES

**G R A D E S**

two and three

**M A T E R I A L S**

paper
pencil
crayons

ometimes the look on a child's face tells you whether an activity is encouraging the right sort of thinking. That's how it was with Chris and COLORED BOXES. When Chris started work on a COLORED BOXES problem, she looked grim and determined. When she solved the problem, an immense smile crossed her face.

The first COLORED BOXES problem I gave Chris was very simple, but it gave her the idea of the game. I handed her two crayons—one black, one red. Then I drew two boxes on a sheet of paper.

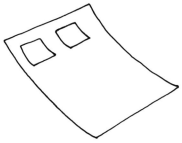

I asked Chris if she could color each box differently.

"Of course I can," she said. She quickly filled one box with red and the other with black. (In a black and white book, we'll have to make do with stripes for red and cross-hatches for black. Does your child want to draw stripes and cross-hatches? Great, try it!)

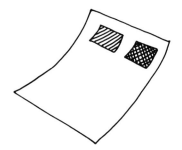

"That's excellent," I responded. "I wonder if you can do a harder problem. This time I'll draw four sets of *double* boxes. Each box can be only one color. But each set of double boxes can be any way you want: one box red and the other black; both boxes red; or both boxes black. Can you color the four sets so that each set is different?"

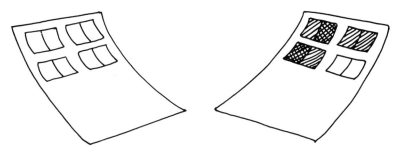

Chris started to color. She thought of three combinations, but got stuck on the fourth. In her search for a solution, she colored a single box half red and half black.

"I'm sorry, Chris, but that's against the rules. You have to fill each single box with a single color. Here, I'll draw a new double box for you."

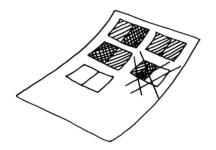

"I can't think of another way to color," said Chris.

"Why don't you start by drawing the first box red. What colors could go in the second box? You could put black in the second box,

but you already have red and black. You could put red and red,

but you already have that combination. Let's see what happens if you start with black. What choices do you have for the second . . ."

Chris interrupted me, "Don't tell me! I know how to do it!" She started to draw, and sure enough, she had it.

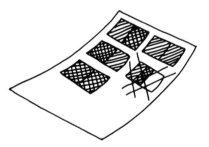

Now that Chris understood the game, I made things still harder. I did this in either of two ways. Sometimes I drew triple- or quadruple-box sets for Chris to color. Other times I gave her three or four different colors of crayons.

When I drew triple boxes and gave Chris two colors, she had to find eight different combinations—that's two to the third power: $2^3 = 8$.

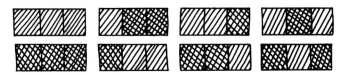

When I drew quadruple boxes and gave her two colors, she had to find sixteen combinations, or two to the fourth power: $2^4 = 16$.

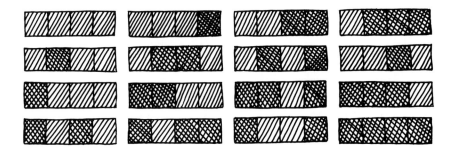

When I drew a single box and gave her three colors, she had to find three "combinations": $3^1 = 3$.

When I drew twin boxes and gave her three colors, she had to find nine combinations: $3^2 = 9$.

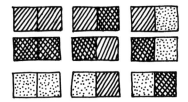

When I drew triple boxes and gave her three colors, she had to find twenty-seven combinations: $3^3 = 27$.

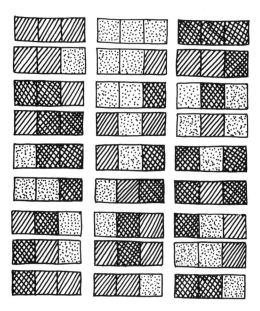

When I drew twin boxes and gave her four colors, she had to find sixteen combinations: $4^2 = 16$.

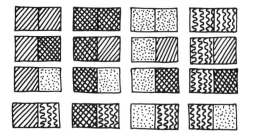

Of course I didn't give Chris all these problems at one sitting. Some days, after solving one or two COLORED BOXES problems, she'd plead for one more. She really loved this drawing game. But I was firm and unrelenting, on the grounds of "Always leave 'em screamin' for more."

What is mathematical about this drawing assignment? The best way to find all possible color combinations is to work in a planned systematic fashion. First, you should find all the combinations that begin with red, then all the combinations that begin with black, then all the combinations that begin with blue. Unless you work systematically, you will get confused, especially when there are lots of boxes. After working out a few COLORED BOXES problems, children appreciate the need to organize their thoughts. They are willing to plan before they draw, and to evaluate their work as they go along. This is the same attitude mathematicians have towards solving problems. When confronted with a new problem, mathematicians develop a systematic, logical, planned evaluation of all possible solutions. By playing COLORED BOXES, therefore, children don't learn math per se. They learn to think like mathematicians.

One day Chris said, "This isn't really a game, because you have to think so hard. But it really is a game, because it's so much fun." Aha! I knew the game was working.

# Appendices

# WHAT ABOUT COMPUTERS?

Ben and I were playing a math game on the computer. In this game, you solve equations and then direct a fly through a maze to pick up the numbers you need to make your answer. The trick is to get the fly to the numbers before it gets gobbled up by an attacking spider. The math problems weren't difficult for Ben. They were basic addition-and-subtraction facts for the numbers one through ten. But he did need to practice these equations. Ben played the game with concentration and delight for about fifteen minutes. He looked forward to playing again on another day. The game was a success.

Jean and I played a different game. In this game animals walk across the computer screen, and when they come to a halt, you count and tell the computer how many elephants or tigers or giraffes you see. Jean liked the animals. They were pretty. But she had a devil of a time counting them, and it was hard for me to help her. All I could do was point to the screen. I couldn't move the animals for her; their movements were already programmed into the software. I couldn't give her the animals to hold; they were only lights on the screen. Jean enjoyed the picture show, but she was frustrated by the problems. This game was not a success.

To play computer games with your child, you have to buy special game software. The software is produced by various companies and is by no means of uniform quality or usefulness. You can't assume that computer games will always be educationally sound. Some games are fine; others are pointless.

The most common variety of software for children is the type that teachers call "drill and practice." Drill and practice on basic arithmetic facts and figures are exactly what you get in this software. Many programs are nothing more than workbook pages done in electronic form. And just like workbook pages, these programs are often boring.

Some programs, however, make imaginative games out of the repetitive drill that is necessary to memorize number facts. Ben and I played a game like that. But here's the rub. Even the clever-est computer games aren't necessarily *better* at helping children memorize facts than old-fashioned homemade games. Educational researchers have looked into this. In head-to-head combat, board games held their own against their high-tech challengers. Since a do-it-yourself board game like NUMBER LADDER costs five cents to make while a computer can cost thousands of dollars, NUMBER LADDER is certainly cost-effective.

A second kind of computer software aims to help children un-derstand mathematical concepts. The animal-counting game that Jean played is an example. Jean wasn't practicing a skill she had already mastered. Instead, she was coming to terms with a new concept—that of naming one number for every object you count. But because she couldn't control the pictures on the screen, the computer was only getting in her way. That's the heart of the computer problem. Children learn best when they have objects they can hold in their hands and push around at whim. A computer elephant may be cute, but it limits the ways a young child can learn. It's different with older children. Older children can do without concrete objects. They can learn from symbols on the screen—at least some of the time.

A third computer activity is programming. To program a com-puter, you have to know a language that the computer understands. If you know this language well enough, you can order the computer to play your own invented games. Many educators claim that learning a computer language increases children's ability to reason logically and think mathematically. A number of university studies have tried to verify these claims. At Kent State University, re-searchers taught LOGO programming to first-graders. LOGO was specially designed for children, and the first-graders in the study received systematic instruction for eighty minutes a week. After just twelve weeks, the children made impressive gains in tests of creative thinking. On the other hand, Bank Street College of Edu-cation taught children in grades three through six how to program in LOGO and found that computer programming failed to live up to the claims of its admirers. Someday somebody will account for these discrepancies. One thing seems certain already, however. Any successful programming project demands systematic instruction—the kind of thing better done by teachers in school than by parents at home.

What conclusions can be drawn from all of this? When it comes to drill and practice, there is some good software and some bad. But even the best software can't outdo a good homemade game. When it comes to teaching concepts, computers have their limits, especially with young children. When it comes to programming, the jury is still out. We just don't know the benefits of teaching computer programming to a generation of children.

Considering the drawbacks to computers, why are so many educators enthusiastic about these machines? There is, in fact, a lot to be enthusiastic about. Computers are relatively recent additions to classroom and home life, and we're just beginning to figure out how to use them to help children learn. New software is developed every day; as software companies start listening to teachers, there's a good chance that excellent programs will find their way onto floppy disks. But I also suspect that some educators are seduced by children's excited responses to computers. Childen do get excited by the color, the beeps, the little creatures, and the action of most computer games. Children are also faddish, and it excites them to be in on the computer whirlwind. Excitement is OK, but it can't be an educator's or a parent's main goal. TV is exciting, too. But watching TV has the unfortunate side effect of deadening many a child's taste for the more subtle excitement of reading books.

All of this is by way of warning: computers are fine, but they're not an educational miracle. If your home is computerless, you needn't worry that you're depriving your child. If you do own a computer, be picky when picking out software. Don't be fooled by a few laser guns and spaceships. Here's a list of software I've found worthy of on-line time. The list goes in order of difficulty from elementary to advanced.

*Mix and Match* (Children's Television Workshop)
*Sticky Bear Numbers* (Weekly Reader Family Software)
*Sticky Bear Shapes* (Weekly Reader Family Software)
*Bumble Games* (The Learning Company)
*Moptown Parade* (The Learning Company)
*Gertrude's Puzzles* (The Learning Company)
*Estimation* (Math and Computer Education Project)
*Muppetville* (Sunburst Communications)
*Math Rabbit* (The Learning Company)
*Number Stumper* (The Learning Company)
*Math Magic* (Mind Play)
*Math Maze* (DesignWare)
*Addition Magician* (The Learning Company)
*Minit/Maxit/Target* (Milliken Publishing Company)
*Adventures in Math* (IBM Software)
*Math Shop* (Scholastic)
*Quotations* (Scholastic)

I believe the best use of computers with young children is literary rather than mathematical. I've seen remarkable things happen when children use simple word-processing programs to help them write stories. Children feel more in control of computerized writing. Even very young authors adjust words, clarify ideas, and insert new thoughts, when such editing requires only the tap of a computer key. Computers free children from concerns over handwriting and spelling—the bugaboos of pencil-and-paper work. There are several good word-processing programs suitable for children on the market. Here are a few I've used and can recommend.

*Writing Assistant* (IBM Software)
*Magic Slate* (Sunburst Communications)
*Bank Street Writer* (Scholastic)

Here are addresses for all the computer companies mentioned above:

The Learning Company, 545 Middlefield Road, Suite 170, Menlo Park, Cal. 94025

Children's Television Workshop, One Lincoln Plaza, New York, N.Y. 10023

Math and Computer Education Project, Lawrence Hall of Science, University of California, Berkeley, Cal. 94720

Weekly Reader Family Software, Xerox Education Publications, 245 Long Hill Road, Middletown, Conn. 06457

Sunburst Communications, 39 Washington Avenue, Pleasantville, N.Y. 10570

DesignWare, 185 Berry Street, San Francisco, Cal. 94107

Milliken Publishing Company, 1100 Research Blvd., P.O. Box 21579, St. Louis, Mo. 63132-0579

Mind Play, 82 Montvale Avenue, Stoneham, Mass. 02180

IBM Software, P.O. Box 1328-C, Boca Raton, Fla. 33432

Scholastic, 730 Broadway, New York, N.Y. 10003

# MATH BOOKS
# TO READ ALOUD

When you read aloud to your child, do you ever choose math books? I mean good storybooks that have mathematical content. You may be familiar with counting books that concentrate on the numbers one to ten, or one to twenty. There are also excellent books that tell about bigger numbers, and about shapes, sizes, time, logic, and patterns. Two important purposes are served when you read math books to your child. Your child learns a little more about math. And your child learns that math belongs everywhere —even at bedtime story hour. Here is a list of math books for reading aloud that are worthy of bedtime—or afternoon or morning-time. The books are listed in order of difficulty, starting with the easiest.

*Who's Counting?* by Nancy Tafuri
*Shapes* by Rosalinda Kightley
*Shapes, Shapes, Shapes* by Tana Hoban
*Counting Wildflowers* by Bruce McMillan
*The Sesame Street 1,2,3 Storybook* by Emily Perl Kingsley and others
*Babar's Counting Book* by Laurent de Brunhoff
*Richard Scarry's Best Counting Book Ever* by Richard Scarry
*A Book of Seasons* by Alice and Martin Provensen
*Farmer Mack Measures His Pig* by Tony Johnston
*What Time Is It, Little Rabbit?* by J. P. Miller
*The Purse* by Kathy Caple
*The Stopwatch* by David Lloyd
*Count on Your Fingers African Style* by Claudia Zaslavsky
*Number Families* by Jane Jonas Srivastava
*Metric Can Be Fun* by Munro Leaf
*Anno's Counting House* by Mitsumasa Anno
*The Biggest, Smallest, Fastest, Tallest Things You Ever Heard Of* by Robert Lopshire

*How Much Is a Million?* by David M. Schwartz

*The Month-Brothers: A Slavic Tale* retold by Samuel Marshak

*All in a Day* by Mitsumasa Anno and others

*Anno's Mysterious Multiplying Jar* by Masaichiro and Mitsumasa Anno

*Giants of Land, Sea, and Air: Past and Present* by David Peters

*Charlie Brown's Super Book of Questions and Answers* edited by Hedda Nussbaum (there are several books in this series)

*Kids Are Natural Cooks* by Roz Ault

*Anno's Hat Tricks* by Akihiro Nozaki and Mitsumasa Anno

*Socrates and the Three Little Pigs* by Tuyosi Mori

*Anno's Math Games* by Mitsumasa Anno

*Math for Smarty Pants* by Marilyn Burns

*The I Hate Mathematics! Book* by Marilyn Burns

*The Kids' Complete Guide to Money* by Kathy S. Kyte

*The Phantom Tollbooth* by Norton Juster

Older children—fourth-graders on up—might enjoy the monthly magazine *Dynamath*. You can write to Scholastic Dynamath, P.O. Box 644, Lyndhurst, N.J. 07071, for subscription information.

# EXPLAINING NUMBER ESP

## Trick with 1,089

You start with the digits A, B, C, where $A > C$. When you use A in the hundreds place, you can represent that as 100A, so the three-digit number can be shown as

$$(100A + 10B + 1C),$$

and its reverse as $\quad (100C + 10B + 1A).$

In part one of the problem, you take the larger three-digit number (the first) and subtract the smaller (the second). When you make the proper place-value exchanges (borrowing from 10B because 1C is smaller than 1A, and so forth), you have

$$[100(A-1) + 10(B+10-1) + 1(C+10)]$$
$$-\ [100(C) \qquad + 10(B) \qquad + 1(A)],$$
$$\overline{100(A-1-C) + 10(B+10-1-B) + 1(C+10-A)}$$

or, more simply (with the Bs canceled out),

$$100(A-1-C) \qquad + 10(9) \qquad + 1(C+10-A).$$

In part two of the problem, you take the solution to part one, reverse the order, and add the two numbers together:

$$[100(A-1-C) \qquad + 10(9) \qquad + 1(C+10-A)]$$
$$+\ [100(C+10-A) \qquad + 10(9) \qquad + 1(A-1-C)]$$
$$\overline{[100(A-1-C)+(C+10-A)]+10(9+9)+1[(C+10-A)+(A-1-C)]}.$$

Rearrange things a bit, and you'll see that in the ones place you have $(C-C) + (A-A) + (10-1)$; the letters cancel out, leaving you with $(10-1)$, or 9. In the tens place you have $(9+9)$, or 18. In the hundreds place you can again cancel out the letters, again leaving you with $(10-1)$, or 9. So the last total is now simplified:

$$100(9) + 10(18) + 1(9)$$

or $\qquad 900\ +\ 180\ +\ 9\ =\ 1089.$

But if you use consecutive digits for A and C, you come up with a different total—because C = (A—1). Therefore, we can replace all Cs with (A—1) and end up with

$$100[(A-1)-(A-1)] + 10(9) + 1(A-1+10-A).$$

The two (A—1)s in the hundreds place cancel out, and so do the two As in the ones place, so we're left with

$$100(0) + 10(9) + 1(9)$$

or 99. Reverse those two digits—you get 99. Add 99 to 99 and you always get 198. As anyone can see!

## Trick with Nines

This trick has to do with the mathematical notion of casting out nines. You start with a number

$$10,000A + 1,000B + 100C + 10D + 1E$$

and rearrange the digits for

$$10,000D + 1,000B + 100C + 10A + 1E.$$

But if you add each set of digits without regard to place value (as when you cast out nines), you get the same answer:

$$(A+B+C+D+E) = X$$
$$(D+B+C+A+E) = X.$$

And $X - X = 0$. Since cast-out-nine for the first number minus cast-out-nine for the second number must equal cast-out-nine for the answer, and since only nine or multiples of nine will cast out to zero, you must get nine or a multiple of nine in answer to

$$(10,000A + 1,000B + 100C + 10D + 1E)$$
$$- (10,000D + 1,000B + 100C + 10A + 1E).$$

# Trick with 6,174

You start with the problem

$$
\begin{array}{r}
(1{,}000A + 100B + 10C + 1D) \\
- (1{,}000D + 100C + 10B + 1A) \\
\hline
= (1{,}000E + 100F + 10G + 1H).
\end{array}
$$

Add the separate digits E+F+G+H—disregarding place value—and you get nine or a multiple of nine (see trick with nines). Rearrange this new set of digits so that EFGH is now a new equation for $(1{,}000A + 100B + 10C + 1D)$, and you end up with a new set for EFGH. But every set of digits EFGH must add up to nine or a multiple of nine. There are many sets of digits that meet this requirement, so it can take a while to come up with 6,174.

Two things help to make the special number come up sooner. In the course of changing EFGH into a new set for $(1{,}000A + 100B + 10C + 1D)$, you always arrange it so that $A > B > C > D$. This limits the number of combinations you go through. Also, every time you subtract you get a completely new set for EFGH—until you end up with the digits 6, 1, 7, 4 in any order whatsoever. Once you have these digits—arranged as 7,641 or 1,467 or any way at all—the next trick manipulation brings you to 6,174. What's more, you are now at the end of the line. Unlike all other sets of four digits, 6,174 turns back into 6,174 in a never-ending cycle.

This strange number was discovered in the 1940s by the Indian mathematician D. R. Kaprekar. There are no two-digit numbers that act like this. There is a three-digit number—495. There are no five-digit numbers and there are probably no numbers of six, seven, eight, or nine digits that act like this—although proving it would keep computers working for quite a while. This means that 6,174 and 495 are, in their own ways, very special indeed.

# A NOTE TO TEACHERS

*Games for Math* can be used in the classroom. Many of the games are ideal for small group use. A math group with six to eight children can play LOTS OF BOXES, BUNCH OF BEANS, or WHAT DID I DO? If you want to help half a math group and don't know what to do with the other children, try assigning games to play. Games like MATH CHECKERS can keep a couple of children busy learning on their own while you attend to the remaining math group members. A two-minute game is a wonderful way to warm up a math group or end math time with a giggle.

A few of the games are suited for the whole class at once: among them, COUNTING AND ESTIMATING, NUMBER TRAILS, and WHAT'S IT WORTH? You will surely find others as you read through the book.

Some of the games work best when played with just one or two children at a time. You might, therefore, arrange for individual children to play games with classroom aides, or recruit parents to spend half an hour in your classroom playing games with individual children. Sixth-graders might act as game leaders for first-, second-, and third-graders. Games like FAST TRACK and DOUBLE IT make wonderful seat-work assignments—a pleasant change from worksheets.

You might encourage parents to use games at home: simply pick out games that suit their children's needs. If a child needs help understanding addition, send home instructions for WHAT DID I DO? If a math group is studying place value, send home instructions for FIFTY WINS. You can also introduce games to parents at conferences or parents' nights. How about an annual game night at school? Invite parents to spend an evening learning games and making game boards to take home. Or try starting a game lending library in your classroom. That way children can sign out math games to play at home for a week.

Games can be used at four different instructional levels: curriculum introduction, co-instruction, drill for mastery, and review.

You can certainly preview material with a game. For instance, STAR COUNT is a sound way to introduce multiplication.

A game makes an excellent co-instructional tool. When you are developing the notion of place value, THROW A NUMBER should prove a valuable addition to your teaching.

Tedious aspects of mathematics, such as drilling to memorize addition and subtraction facts, can be made palatable by playing NUMBER LADDER, or similar games.

Finally, games are a wonderful way to review old topics in math. In December your students need to review aspects of numeration covered in September. Well, then, play GRASSHOPPER.

Many of these games call on children to handle objects or draw pictures rather than deal with numbers symbolically. Is this wise? Ninety percent of the research studies on the use of manipulatives in the math curriculum show that children do as well or better in math when they use manipulative materials. Regrettably, other studies indicate that thirty-seven percent of kindergarteners through sixth-graders use manipulative materials less than once a week, and almost ten percent of the same group *never* use manipulatives in school. Perhaps overwhelmed teachers lack simple, pleasant, non-time-consuming ways to integrate manipulative materials in the classroom setting. *Games for Math* can help.

You should find it easy to integrate *Games for Math* into your mathematics curriculum. In Piagetian terms, Part One of the book is geared to pre-operational children. Most kindergarteners and many first-graders are pre-operational. As you know, these children can do quite a lot of math, as long as their special brand of numerical logic is taken into consideration.

Recent investigations show how heavily young children rely on counting to understand numbers. In fact, children go through three stages when learning to count. Stage one is rote counting. At this stage children learn to say the proper number words in a fixed order. Stage two is rational counting. At this level children appreciate the need to name one and only one number for every object counted; they know that the order of counting objects doesn't affect the count; they appreciate that the last number counted quantifies the whole group. Stage three comes when children conserve number and appreciate the lasting equivalency of whatever they count. Pre-operational children can only master the first two stages of counting. The games in *Chapter One: Counting Counts* are designed to help children master these number skills.

Educators know we can't teach children to move from pre-operational logic to the logic of concrete operations. We can, however, provide pre-operational children with an environment that encourages experimentation with intellectual ideas. That's the point of the activities in *Chapter Two: Thoughts to Grow On.*

Just as pre-operational children can count before they conserve number, they can measure before they conserve length. At least they can if measurement is presented in the context of a proper pedagogical environment. These children can also learn a bit about common geometric shapes. In *Chapter Three: Size and Shape* you'll find some informal measurement activities appropriate for young children, along with a couple of geometric games.

Sometime in first grade most children reach the stage of concrete operations. These children can conserve number. They are ready, therefore, to expand their mathematical explorations. They are ready to play the games in Part Two of *Games for Math.*

*Chapter Four: Fancy Counting* gets children counting large numbers, skip counting, and counting on from numbers other than one. *Chapter Five: Adding and Subtracting* has a dual role. Some of the games help children understand the meaning of addition and subtraction. Other games make playful the drill required to memorize addition and subtraction facts.

Sometime in first grade most children develop the logic required to conserve length. (Conservation of number usually precedes conservation of length by a bit.) These children can proceed to more formal measurement activities than their pre-operational friends: thus *Chapter Six: Size and Shape II.* It is best to start your measurement curriculum with nonstandard units of measure.

Measuring a book with toothpicks should precede measuring a book with centimeters. When the children in your class are familiar with the concept of measurement developed by using nonstandard systems, they are ready to sew, cook, and play the measurement guessing game in *Size and Shape II*.

Once children can conserve number, they can think about different kinds of numerical groupings with equal ease. These children can understand multiplicative concepts of numbers as well as additive concepts. That's why you can present some multiplication and division to first-graders and they will know what you are talking about. Beginners shouldn't work with symbolic notation. They should, however, investigate multiplication and division informally, and with concrete objects. This will prepare them for formal and symbolic representations in later grades. *Chapter Seven: Multiplication and Division* has three games (STAR COUNT, VICTOR VAMPIRE'S BIRTHDAY, and LOTS OF BOXES) perfectly suited to first-, second-, and third-graders, who are just starting to understand these arithmetic operations. There is also a game (CALCULATING MATH) to help third-graders drill multiplication facts. Some of the games in *Chapter Five* (for instance, MATH CHECKERS and NUMBER LADDER), can be transformed quite easily into multiplication and division games.

*Chapter Eight: The Number System* deals with place value and the base ten quality of our number system. Base ten and place value are mathematically the most sophisticated and important topics tackled by elementary school students. Because base ten patterns are deceptively easy to follow, we sometimes think children have the idea even when they do not. That's why it's a good idea to continually review these ideas throughout the elementary school years—even when it appears that children have mastered this area of mathematics. Such careful attention to the number system pays off later when children grapple with long division, decimals, and many other areas of higher mathematics.

*Chapter Nine: The Bigger the Better* is the most arithmetically challenging section of the book. Before children can play these games, they must feel comfortable with multidigit addition and subtraction, in which regrouping is part of the process.

Part Three is composed of two chapters, *Strategy Games* and *It's a Puzzle.* Both chapters, in nonroutine ways, help children develop skills in mathematical problem solving.

A note on mathematical language: mathematicians know the importance of distinguishing between the words *number* and *numeral.* Elementary school children and many of their parents, however, are only confused by such terminology. That's why I chose to use the word *number* throughout the book, even when *numeral* was the more exact term.

*Games for Math* is one good source for classroom math games, but there are other resources available to teachers. *The Arithmetic Teacher*, a monthly journal published by the National Council of Teachers of Mathematics, 1906 Association Drive, Reston, Va. 22091, is an excellent magazine. It is filled with valuable teaching suggestions along with a bit of easy-to-understand educational theory and current research discoveries.

The following companies all distribute useful math materials and math books. You can write for catalogs.

Dale Seymore Co., P.O. Box 10888, Palo Alto, Calif. 94303
Activity Resources, P.O. Box 4875, Hayward, Cal. 94540
Creative Publications, 5005 West 110 Street, Oak Lawn, Ill. 60453
Didax, 5 Fourth Street, Peabody, Mass. 01960
Cuisenaire Co., 12 Church Street, Box D, New Rochelle, N.Y. 10802

The following magazines for elementary-school teachers may provide you with monthly inspiration for your math curriculum:

*Teaching K–8* (formerly *Early Years*), P.O. Box 912, Farmingdale, N.Y. 11737-9612
*Learning*, P.O. Box 2580, Boulder Col. 80322
*Instructor*, P.O. Box 6099, Duluth, Minn. 55806-9799
A few games now and again can make the year fly by.